COLOUR DICTIONARY OF

GARDEN PLANTS

MARSHALL CAVENDISH

This edition published in 1994 by
Marshall Cavendish Books, London

Editor: Susanne Mitchell
Art editor: Caroline Dewing
Designer: Sheila Volpe

ISBN 1 85435 726 3

British Library Cataloguing in
Publication Data:
A catalogue record for this book is
available from the British Library

Printed and bound in Malaysia

Contents

Introduction

This illustrated dictionary is more than a list of plant names. As well as being a reference book of popular plants, it is intended to be a practical help for anyone who wants to get the best from a flower border. Although it has not been possible to list all the species, let alone varieties, in a book this size, the general advice given for the most popular varieties of a species will generally also apply to others. So unless you buy a rather unusual plant, you will find it within these pages.

The title of this book is *Colour Dictionary of Garden Plants*. The vast majority of the plants included are what gardeners traditionally call herbaceous border plants—those which die down in the autumn and reappear in the spring. There are also flowering perennials which do not die down and lose all sign of growth and foliage, which some gardeners call 'border perennials'. However, to confine one's flower borders to plants conforming to a narrow botanical definition is to miss many opportunities—there is no reason why you

Mixed planting of herbaceous flowers

should not mix herbaceous plants with suitable shrubs, and even in a more traditional herbaceous border it would be a pity to exclude biennials such as foxgloves and bulbs such as lilies. We have therefore included a small selection of biennials and bulbous plants to complete the flower border.

For reasons of space it has not been possible to include hardy and half-hardy annuals. These are usually grown separately anyway, but do not overlook the contribution they can make as gap-fillers in the herbaceous or mixed border, especially in the early years while the perennial plants are becoming established.

Throughout the book, you will find seasons referred to rather than months, which vary from country to country. As a guide, the seasons are taken to mean the following in the northern and southern hemispheres respectively:

Early spring: March/September
Mid spring: April/October
Late spring: May/November
Early summer: June/December
Mid summer: July/January
Late summer: August/February
Early autumn: September/March
Mid autumn: October/April
Late autumn: November/May
Early winter: December/June
Mid winter: January/July
Late winter: February/August

Using the book

The first half of the book—covering a wide range of plants—has been arranged alphabetically by Latin name. But widely used common names are also included, so if you know a plant only by its common name you should use the common name index on page 119 which will direct you to the relevant page.

To help you decide at a glance whether a plant is likely to be suitable for the soil and aspect that you can provide, key requirements have been presented as symbols for each entry. These are explained on the page opposite.

Herbaceous plants blend well with shrubs

Terms and techniques

We have tried to make this book as practical as possible, so that you can get the most out of the information. If you are an experienced gardener you will probably find everything that you need in the plants' section, but to make the book as useful as possible to beginner and experienced alike, a section at the back of the book is devoted to an explanation of the terms and techniques mentioned here or that you might encounter elsewhere. If you feel that you need to ask about acid soil, or have wondered what is meant by 'whorls', you will find a concise explanation that should enable you to understand the term and so get the best from the book.

Techniques are as important as terms, and practical jobs like propagation are also explained, so even if you've never taken a root cutting or layered a carnation this should be no problem.

What the symbols mean

H	Herbaceous plant that usually dies down for the winter. Occasionally some kinds may be evergreen, or nearly so depending on the weather, and this is mentioned in the plant description where applicable.
(CL)	Climber.
B	Bulb (the term is used loosely and includes corms and tubers).
AD	Does best on acid soil. Avoid alkaline soil.
AK	Does best on alkaline soil. Avoid acid soil.
O	Will grow well on a wide range of soils. However, few plants tolerate water-logged conditions.
(D)	Suitable for dry soil.
(M)	Suitable for moist soil.
FS	Full sun for most of the day.
PS	Partial shade, or sun for only part of the day.
SD	Shade.

The Plants

Acanthus

H/O/FS–PS

The name comes from the Greek word *akanthos* (prickle), alluding to the spiny leaves of some species. These plants were well known to the Greeks and Romans, the leaf form of *A. mollis* being used for the decoration of the Corinthian column. These distinctive plants are effective as isolated specimens as well as being important members of the herbaceous border. In fact they are seen to best advantage as specimen plants where their 'architectural' shape can be appreciated more readily.

How to grow
Best in partial shade, though they will still thrive in full sun. They like moisture but the soil needs to be well drained and not waterlogged.

Cut the stems almost to ground level after flowering. Do not lift or disturb the plants unless really necessary.

Propagation
Seed can be sown in containers in a cold frame in early spring. Prick out seedlings into individual pots, harden off in mid summer, and plunge outdoors. Plant in autumn.

Root cuttings, about 5–6 cm (2–2½ in) long, can be taken in late winter. Root in containers in a cold greenhouse or frame. Pot up singly into small pots when new growth is seen, then treat as seedlings.

Acanthus mollis

Division in mid autumn or early spring is very easy and the usual method of propagation.

SOME POPULAR SPECIES	
A. longifolius (Dalmatia, Balkans) Dark green, wavy, deeply-cut leaves about 45 cm (1½ ft) long. Lilac flowers in early and mid summer. Freer-flowering than *A. mollis*. 75 cm (2½ ft). **A. mollis** (Italy) (Bear's breeches) Shiny green leaves with scalloped edges. White	and purple flowers, shaped like dragon-heads, on bold spikes, all summer. 90 cm (3 ft). **A. spinosus** (South East Europe) Horizontally radiating strap-like leaves with long, stiff, spines. Stiff spikes of purple and white flowers guarded by sharp spines. 90–120 cm (3–4 ft).

Achillea

Yarrow, milfoil
H/O/FS

Named after Achilles, the Greek hero, who is said to have first used the plant medicinally. There are about 200 species, the dwarfer ones suitable for the rock garden, the taller ones for the herbaceous border. Some are coarse and weedy, others are good garden plants. All make good cut flowers.

How to grow
Easy to grow plants, that thrive best on stony gravels and chalk soils, but grow well on almost any soil. A sunny position is important.

Achillea filipendulina 'Gold Plate'

Propagation

Seed can be sown in containers in early spring, although named varieties may not come true to type. Sow on the surface and germinate in a cold frame or warm greenhouse. Prick off seedlings into individual pots, harden off at the four to six leaf stage, and plunge outdoors. Protect during the winter, harden off and plant out in spring.

Division is easy, and best done in spring, although in mild areas autumn is also suitable.

SOME POPULAR SPECIES	
A. filipendulina *(A. eupatorium)* (Caucasus) Flattened, bright yellow flower heads, mid summer to early autumn. Light green, feathery leaves. The true species is not much grown as there are improved varieties such as 'Coronation Gold' (greyish leaves) and 'Gold Plate' (lemon). 90–150 cm (3–5 ft). A. millefolium 'Cerise Queen' (Europe) (Yarrow) Flat heads of cherry-red flowers, all	summer. The true species is a lawn weed. 60–75 cm (2–2½ ft). A, ptarmica 'The Pearl' (Europe) (Sneezewort) The true species is seldom used, this double variety with white pompon 'buttons' being the most popular. Flowers mid summer to early autumn. 60 cm (2 ft). A. taygetea (Middle East) Silvery-grey pinnate leaves. Flat heads of pale yellow flowers, all summer. 45 cm (1½ ft).

Acidanthera

B/O/FS

When the plant listed here was first introduced it was thought to be a gladiolus, and it was given an award by the British Gladiolus Society under the name *Gladiolus murielae*. Soon afterwards it was decided that the plant was really an acidanthera, and is now grown as *A. bicolor murielae*. The name acidanthera comes from the Greek *akis* (a point) and *anthera* (an anther).

Acidanthera bicolor murielae

How to grow

Start off in pots from mid winter onwards, or plant outside in late spring. Keep frost-free. Plant the corms 8–10 cm (3–4 in) deep. The plants do best in well-drained rocky or gritty soil, but should be well watered while growing. Full sun is important, and it is most reliable in mild areas. Lift in the autumn and store the corms in a frost-free place—ideally in the range 7–13°C (45–55°F).

Propagation

Mature corms produce one or more cormlets during the season. Separate these when lifting in autumn. Pot up eight to ten cormlets in each 18 cm (7 in) pot in early spring, setting them 3 cm (1½ in) deep. Grow on in a cold frame, then in autumn remove the corms and overwinter in trays as before. Pot up again in early spring and repeat the cycle for two more seasons, by which time the corms should have reached flowering size.

SOME POPULAR SPECIES	
A. bicolor murielae (Abyssinia) Sword-shaped leaves. Fragrant, star-like, spaced	out, pure white flowers with a deep maroon blotch in the centre. 75 cm (2½ ft).

Aconitum

Monkshood
H/O/FS–PS

Aconitum is an ancient Greek name for a poisonous plant, *akon* (a dart) probably indicating that arrows were at one time poisoned with the juices of the plant. All parts of the plant are poisonous. The common name refers to the helmet-shaped individual flowers.

How to grow
Semi-shade and moist but not waterlogged soil suit them best, but a sunny spot is acceptable provided the soil is not too dry. Mulch each spring.

A. *napellus* will usually flower again later if dead-headed promptly. Cut back all stems in autumn.

Propagation
Seed can be used, but you cannot depend on the plants to come exactly true to type. Sow in containers in a cold frame in spring. Prick out into individual pots and harden off in late spring or early summer. Plunge outdoors for the summer, protect for the winter, then plant out the following spring.

Division is a more reliable method, and very easy. This can be done in autumn or spring. Make sure each segment has several eyes.

Aconitum fischeri

SOME POPULAR SPECIES	
A. x arendsii (Garden origin) Strong spikes of blue hooded flowers produced in late summer. 1.2 m (4 ft). **A. fischeri** *(A. carmichaelii)* (China) Violet-blue flowers, late summer. 90 cm (3 ft). **A. napellus** (Europe, Asia)	Deeply cut dark green leaves. Deep blue flowers with high hoods, mid and late summer. Also pink or white varieties. 60–90 cm (2–3 ft). **A. wilsonii** (Eastern China) Tall spikes of violet-blue hooded flowers, from early to late autumn. 1.2–1.8 m (4–6 ft).

Adonis

H/O(M)/FS–PS

A genus of about 20 herbaceous perennials and hardy annuals. Named after Adonis, one of Venus's lovers, whose blood is supposed to have stained the petals of the pheasant's-eye adonis (*A. autumnalis*) after he was wounded by a wild animal. The herbaceous plants described here are suitable for the front of the border or even for a rock garden. They are especially useful because they flower so early.

Adonis vernalis

How to grow

Best in rich, moist but well-drained soil. The earliest flowers may need protection from the weather. Plant in late summer or early autumn, setting the crown 2.5 cm (1 in) below soil-level. The plants disappear below ground by mid summer.

Propagation

Seed-raising is sometimes slow, and germination is erratic, but sowing in containers under a cold frame in mid or late winter and then bringing into warmth (16°C/60°F) in early spring, can give good results. Prick out into individual pots and grow on in the cold frame until the following spring, then pot on.

Protect for one more winter, then plant out in spring.

Division is easier and quicker. It can be done in spring, but early or mid autumn is a better time.

SOME POPULAR SPECIES	
A. amurensis (Manchuria, Japan) Much-divided, almost fern-like foliage. Yellow flowers in late winter and early spring. 'Flore-pleno', with double flowers, is the form usually grown. 30 cm (1 ft).	**A. vernalis** (Europe, Siberia) Finely dissected, ferny leaves. Plants covered with bright yellow single flowers about 5 cm (2 in) across in early spring, best in sun. 23 cm (9 in).

Agapanthus
African lily, lily of the Nile
H/O/FS

A group of hardy and half-hardy perennials with thick, strap-like leaves, which may be evergreen. The genus takes its name from two Greek words: *agape* (love) and *anthos* (flower). Although good border plants in suitable areas, most of them will have to be given protection in winter and therefore tend to be better as tub plants.

How to grow

A sunny, sheltered position is best. The hardier species can be planted outdoors in mid spring, setting the crowns 5 cm (2 in) below soil-level. Protect with about 15 cm (6 in) of bracken, coarse sand, or pulverised bark from mid autumn to mid spring but even then there may be losses in very severe winters. In milder areas, however, they can be very successful left outdoors.

Cut the flower stems to ground level once they have finished blooming.

Propagation

Seedlings will take two or three years to reach flowering size but can be germinated quite easily by sowing in warmth (13–18°C/55–65°F) in mid spring. Prick out half a dozen seedlings into each 13 cm (5 in) pot, and grow on a bit cooler. During the summer, pot up individually into 9 cm (3½ in) pots, and overwinter in a minimum 7°C (45°F). Pot on in spring, harden off and plunge outdoors for the summer. Overwinter in a frost-free place again, then harden off in spring before planting out.

Division of flowering-sized clumps in mid spring is the easiest and quickest method.

SOME POPULAR SPECIES	
A. africanus (*A. umbellatus*) (Cape Province) Umbels of deep blue flowers, mid and late summer. Evergreen, but not really hardy. 60–75 cm (2–2½ ft).	**A. 'Headbourne Hybrids'** (Garden origin) Spherical umbels of flowers from pale blue to deep violet-blue. Among the hardiest agapanthus. 60–75 cm (2–2½ ft).

Agapanthus needs a sunny sheltered border

Ajuga
Bugle
H/O/PS

A genus of about 40 hardy annuals and herbaceous perennials. The species widely grown are creeping perennials useful as a ground cover or for the front of the border.

Ajuga reptans 'Variegata'

They are, however, perhaps too aggressive for growing among choice plants. The name could be a corruption of the Latin *abiga* (a plant used in medicine), or from *a* (no) and *zugon* (a yoke) alluding to the calyx lobes being equal.

How to grow
A. genevensis will do better in a sunny, well-drained site, but the other species, including the popular *A. reptans,* are best in moist soil in partial shade. Very easy plants to grow.

Propagation
Divide large clumps at almost any time, but preferably in spring or summer (just before or after flowering).

SOME POPULAR SPECIES	
A. genevensis (Europe) Coarsely-toothed, ground-hugging leaves. Blue, pink, or white flowers carried on short spikes, in early and mid summer. Not rampant. 15–30 cm (6–12 in). **A. pyramidalis** (Europe) Dense spikes of brilliant gentian-blue flowers, late spring to early summer. Not as rampant as *A. reptans*.	**A. reptans** (Europe) (Bugle) Creeping plant with short upright spikes of blue, white, or pink flowers in late spring and early summer. There are varieties with decorative foliage. Leaf colours include reddish-purple ('Purpurea'), mottled bronze, pink and yellow ('Multicolor', syn. 'Rainbow') and grey-green and cream ('Variegata').

Alchemilla
Lady's mantle
H/O/FS–PS

From the Arabic word *alkemelych*, alluding to the plant's use in alchemy. Individual flowers are not colourful or large, and the whole heads of most species are greenish—but they have a feathery gracefulness that makes them very popular. The leaves, too, are attractive, those of *A. mollis* being popular with flower arrangers.

Alchemilla mollis

How to grow

Provided they have a position in sun, or even better partial shade, the alchemillas need little attention. They do best on a moist but well drained soil. Cut the stems back to within a couple of inches of the ground after flowering.

Propagation

Seed provides a ready means of increase. It can be sown in containers under a cold frame in early summer, but better results come from sowing in warmth (16–21°C/60–70°F) in spring. Seedlings are usually ready to prick out into small pots after about a month—and can be hardened off and plunged in a cold frame in late spring or early summer. Remove the frame lights (tops) completely in mid summer. Plant out in autumn.

Crowns can be divided in early or mid spring, and this is probably the best method if only a few plants are required.

SOME POPULAR SPECIES	
A. alpina (Europe, Greenland, Northern Asia) Silvery-grey finely divided leaves. Sprays of buff flowers in summer. 15 cm (6 in). **A. mollis** (Carpathians to Western Asia)	Kidney-shaped, wavy-edged, grey-green leaves. Dainty sprays of small yellowish-green flowers in early and mid summer. Good ground cover. 30–45 cm (1–1½ ft).

Allium

B/O/FS

Bulbous plants, including many popular vegetables such as chives, garlic, and onions. Although most have an onion smell, some species are very ornamental. *Allium* was the Latin name for garlic, and probably derives from the Celtic word *all* (pungent or burning).

How to grow

An open, well-drained soil in full sun is ideal. Otherwise they will look after themselves. It is best to leave clumps undisturbed for as long as possible—only lift and divide when flowering performance seems to be suffering.

Propagation

Seed is a useful, but slow, method of propagation. Sow thinly in containers of gritty compost in a cold frame in early or mid spring. Do not disturb any of the seedlings that germinate until the following spring, then prick out into individual small pots. Harden off before plunging outdoors in late spring, and grow on until autumn, when they can be given frame protection for the winter. Harden off and plunge again the following spring, then plant out in early autumn.

Division is much quicker and easier. Lift, divide and replant immediately in early autumn. At the same time small offsets can be separated and potted up to grow on for a year or two under a frame.

SOME POPULAR SPECIES	
A. aflatunense (Central Asia) Strap-shaped. Round heads of purple-lilac star-shaped flowers in late spring. 75 cm (2½ ft). **A. albopilosum** (*A. christophii*) (Middle East) Strap-shaped leaves and large rounded heads of small star-like lilac-pink flowers in early summer. 45 cm (1½ ft). **A. cernuum** (USA) Nodding, soft mauve	flowers. Also rose, purple, or white. Summer. 45–60 cm (1½–2 ft). **A giganteum** (Himalayas) Strap-shaped leaves. Deep lilac flowers in umbels about 10 cm (4 in) wide, early summer. 1.2 m (4 ft). **A. moly** (Southern Europe) Tulip-like leaves. Bright yellow star-like flowers, early and mid summer. 30 cm (1 ft).

Allium moly

Alstroemeria
Peruvian lily
H/O(D)/FS

Alstroemeria Ligtu Hybrid

A group of perennials with fleshy, tuberous roots. Named in honour of Baron Clas Alstroemer, a Swedish botanist (1736–94), and friend of Linnaeus. Not dependably hardy in cold areas, slow to establish, and rather sparse for an ideal border plant—but very good for cutting.

How to grow
It is best to obtain pot-grown plants as alstroemerias resent root disturbance. Plant in groups (individual plants will look lost) 10–15 cm (4–6 in) deep, in well-drained soil.

Twiggy sticks may be necessary for support. Do not cut too many stems for the house until the clumps are established. Dead-head the plants unless seed is required, and cut the stems to the ground in autumn once the leaves yellow. *A. aurantiaca* is hardy, but all the others should be protected with a thick layer of peat, bracken, coarse sand, or pulverised bark.

Propagation
Seed usually presents something of a challenge. Try soaking in tepid water for 12 hours before sowing. Maintain 24°C (75°F) for about a month, then reduce the temperature to 10–13°C (50–55°F). Seeds should then germinate in about a month. When large enough, prick out into individual pots and move to a cold frame, plunging the pots in moist peat. Overwinter in the frame, protecting the frame with matting in cold weather. Harden off and plant outdoors.

Division is much easier, but disturb and break the roots as little as possible. Pot up the segments into 15 cm (6 in) pots and grow on in a cold frame for a year before hardening off and planting out.

SOME POPULAR SPECIES	
A. aurantiaca (Chile) Lanceolate leaves on tall, upright stems, topped by the trumpet-shaped flowers in shades of yellow and orange. The hardiest species. 90 cm (3 ft). **A. ligtu Hybrids** (Garden origin) Similar, but good colour	range that includes shades of pink, golden-orange, and flame. 60–75 cm (2–2½ ft). **A. pulchella** (Brazil) Red flowers tipped green. Less hardy than the previous species. 90 cm (3 ft).

Althaea
Hollyhock
H/O/FS

The name *Althaea* comes from the Greek *althaia* (a healing medium), referring to the medicinal use of some of these plants. In fact the constitution of the hollyhock itself is not good—it falls easy prey to rust disease if treated as a perennial. For that reason it is more usual to grow them as biennials, raising fresh plants each year.

Althaea rosea

How to grow

Hollyhocks do well on heavy soils, but need plenty of water in dry periods. Mulch well in spring. Tall plants, or shorter ones in an exposed position, will need adequate staking.

Although dead-heading is not usually necessary for the sake of the plant, it will reduce the number of self-sown seedlings, if these are usually a nuisance.

If you want to retain old plants from year to year, cut them down to 15 cm (6 in) above soil-level after flowering.

Propagation

Seed is best sown in containers under a cold frame in late spring or early summer. The seedlings are usually ready to prick out into 10–13 cm (4–5 in) pots about a month later. Plunge the pots outdoors after a further three weeks, and plant out in early autumn.

You can get some hollyhock varieties to flower well in the first year by sowing them in warmth (13–16°C/55–60°F) in late winter. Prick them out singly into 10 cm (4 in) pots and grow on in slight warmth. Harden off and plant out in late spring.

Cuttings of 10 cm (4 in) basal shoots can be rooted in a cold frame in mid or late spring, but cuttings are more prone to rust disease than seed-raised plants.

SOME POPULAR SPECIES	
A. rosea (China) Well-known tall flower spikes from mid summer to early autumn. There are single and double varieties, in shades of pink, yellow, red and	white. There are dwarf varieties treated as annuals little more than 60 cm (2 ft) high, but the traditional types grown as biennials reach 2.7 m (9 ft).

Alyssum

Madwort
H/O(D)/FS

From the Greek *a* (not) and *lyssa* (madness). Alyssum was once considered to be a remedy for a bite by a mad dog—which gives rise to one of the common names: madwort. The species here are not to be confused with sweet alyssum; they are sometimes used as rock plants but their vigorous habit makes them just as suitable for the front of a border.

How to grow

Easily grown plants, thriving in a sun-baked site that may be too dry for most plants.

To keep the plants compact and neat, cut them back hard after flowering.

Propagation

Easily raised from seed, though for particular varieties you may need to use cuttings. Sow seed on surface of containers in a cold frame, in late spring. Prick out seedlings into individual small pots, then plunge outdoors about three weeks later. Give winter protection. then plant out in late spring.

Cuttings can be taken in late spring or early summer, but they can be tricky. Insert 8 cm (3 in) long cuttings of non-flowering shoots around the edges of 13 cm (5 in) pots under a cold frame. Pot up rooted cuttings singly, and plunge the pots outdoors for the summer. Give winter protection, then plant out.

SOME POPULAR SPECIES	
A. montanum (Europe) Sub-shrubby plant forming a low grey mound topped with loose sprays of lemon-yellow flowers from mid spring to early summer. 10 cm (4 in). **A. saxatile** (Eastern Europe, Russia)	(Gold dust) Lanceolate grey-green foliage. Masses of small yellow flowers from mid spring to early summer. There are several varieties in varying shades of yellow, and a double variety. 15–30 cm (6–12 in.)

Alyssum saxatile 'Citrinum'

Amaryllis
Belladonna lily
B/O/FS

A genus with just the one species. It is named after a shepherdess in Greek and Latin poetry. The specific name *belladonna* comes from the Italian *bella* (pretty) and *donna* (lady), an extract from this plant being used to brighten the eyes.

These beautiful plants are unfortunately not really hardy and are only suitable for growing outdoors in mild areas, and ideally near a warm wall.

How to grow
Plant in early or mid summer in well-drained soil enriched with plenty of rotted compost or manure. Cover the top of the large bulb with about 8–15 cm (3–6 in) of soil. A sunny position near a warm wall is ideal, but they can be very effective in the general border in favourable areas. Mulch each spring, and water and feed regularly during the summer.

Dead-head as individual blooms die, and cut the stem down when it has finished.

Propagation
Seed is slow and can take up to eight years to flower. This makes it a job for the specialist.

Amaryllis belladonna

Division of established clumps is quicker and easier. Lift the bulbs carefully in early or mid summer, when the leaves wither. Separate the offsets and replant immediately or pot up into 12–15 cm (5–6 in) pots to grow under glass for a year before planting out. The offsets will take about three years to flower. Replant the parent bulbs immediately.

SOME POPULAR SPECIES	
A. belladonna (South and South West Africa) Clusters of bright rose-pink flowers, each bloom about 10 cm (4 in) across, atop a stiff, bare stem in	early and mid autumn. The strap-shaped leaves appear from early spring to mid summer. 60–75 cm (2–2½ ft).

Anaphalis
Pearl everlasting
H/O/FS–PS

A genus of attractive grey-leaved plants with white 'everlasting' flowers, popular for cutting fresh or for dried flower arrangements. The name is said to be based on an old Greek word.

How to grow
Well-drained soil and a sunny position are the basic requirements, although they will grow satisfactorily in partial shade.

The plants are likely to become untidy with the advancing season, in which case they can be cut back hard.

Propagation
Seed can be sown under glass in spring.

Anaphalis triplinervis

Germination is likely to take between one and two months at 13–16°C (55–60°F). Prick out into small pots and grow on under a cold frame until the following spring, then pot on. Harden off in late spring, and plunge outdoors for another year, giving winter protection. Plant out in mid or late spring.

Cuttings taken about mid spring will root under a cold frame. Use basal cuttings about 5–8 cm (2–3 in) long. Pot up singly when rooted and grow on under the frame until the following spring, then treat as seed-raised plants.

Division is easier and quicker. Divide mature clumps in early autumn or mid spring.

SOME POPULAR SPECIES	
A. margaritacea (North America, Eastern Asia) Woolly or downy stems and leaves. Heads of tiny pearly-white flowers in late summer. 30–45 cm (1–1½ ft). **A. triplinervis** (Himalayas) Bold, silvery, ovate leaves, the undersides covered with woolly hairs. Clusters of creamy-white	'everlasting' blooms in mid and late summer. 23–30 cm (9–12 in). **A. yedoensis** (Japan) Lanceolate grey-green leaves and bunches of white flowers in heads about 8 cm (3 in) across, in mid and late summer. This plant is now regarded as a variety of *A. margaritacea*. 60 cm (2 ft).

Anchusa

Alkanet
H/O/FS

For a pure, rich blue colour, the anchusas are indispensable plants. The genus takes its name from *anchousa*, the Greek word for a cosmetic paint used to stain the skin, which used to be made from *A. tinctoria*. The genus includes some biennials (some of them treated as annuals) as well as herbaceous perennials.

How to grow
Anchusas do not stand the winter well in cold, wet districts, so planting is best done in early spring.

They are best planted in groups for a bold effect. They will probably need staking.

Cutting down the plants after flowering often produces another spate of bloom in mid or late autumn.

Propagation
A. azurea can be sown in containers under a cold frame from mid spring to early summer. The seedlings are usually ready to be pricked off into small pots in about a month. Grow on for another three weeks, then plunge outdoors. Plant out in early autumn. *A. capensis* can also be treated the same way, but it can be sown in mid or late winter to bloom the same year. Germinate at 13–16°C (55–60°F), prick out into individual pots and plant out in late spring after hardening off.

Root cuttings of *A. azurea* will root fairly easily in a cold frame if taken in late winter. Make them 6 cm (2½ in) long. They should be ready for potting up in mid or late spring. Harden off and plunge outdoors until planting time in early autumn or the following spring.

The easiest method for *A. azurea* is division of established clumps.

SOME POPULAR SPECIES	
A. azurea (*A. italica*) (Caucasus) Mid green lanceolate leaves and a loose pyramid of bright blue flowers all summer. There are several named selections with flowers of various shades of blue. 1–1.5 m (3–5 ft).	**A. capensis** (South Africa) A biennial often grown as an annual. If overwintered outdoors it will probably need cloche protection. Small blue flowers all summer. Bushy habit. 45 cm (1½ ft).

Anchusa azurea 'Loddon Royalist'

Anemone
Windflower
H/AK/FS–PS

The name is derived from the Greek words *anemos* (wind) and *mone* (habitation), because some species are found in windy places. The species most widely grown has been the subject of several name changes and this may lead to some confusion (although as it is really the hybrids derived from the species that we grow it does not really matter). You may find them listed as *A. japonica*, *A. hupehensis*, or *A.* x *hybrida*, though the first two are actually also distinct species. Let none of this detract from the merits of these invaluable late-flowering plants.

How to grow
Best in a moisture-retentive but not waterlogged soil. Although they thrive on chalky soil, they will also do well on moist soils. Sunshine is endured, but they are also especially good in partial shade.

Do not expect too much from them for the first year. The clumps need to be left undisturbed for several years to make a good show.

Cut the stems down to ground level after flowering.

Propagation
Seed can be sown under glass in early spring, but germination is inclined to be erratic. If you try seed, the plants should be ready for final planting in about 18 months.

Root cuttings about 8 cm (3 in) long can be taken in mid winter and rooted under a cold frame. Pot them singly into 13 cm (5 in) pots in mid or late spring. Harden off, plunge outdoors for the summer, and plant out in early autumn.

Division is easy. It can be done in mid autumn, but is better left until early spring. Replant pieces of crown with well-rotted segments each with two or more buds or emerging shoots.

SOME POPULAR SPECIES

A. x hybrida
(A. japonica)
(Garden origin)
The flowers, in shades of pink and white, about 5 cm (2 in) across, are carried on graceful wiry stems well above a compact mass of foliage. There are good named varieties, flowering late summer to late autumn. 60–90 cm (2–3 ft).

A. x lesseri (Garden origin)
Rose-red flowers carried on erect stems. Other colours may be found, but these are rare. Late spring, early summer. 45 cm (1½ ft).

Anemone japonica

Anthemis

H/O(D)/FS

Anthemis comes from the Greek word given to chamomile. The word *anthemon* (flower) probably refers to the free-flowering nature of these plants.

The chamomile itself, included here as *A. nobilis*, is now technically *Chamaemelum nobile*, but you are not likely to find it sold under that name.

The genus contains about 200 species, including hardy annuals, biennials, and herbaceous perennials. The plants included here need no staking, and the marguerite-like flowers are excellent for cutting.

How to grow

Well-drained soil and a sunny position are the basic requirements.

Propagation

Seed can be sown in containers in a cold frame in late spring or early summer, but for particular varieties cuttings are necessary. Seeds germinate quickly and should be ready to prick out after about a month. Plunge outdoors three weeks later. Protect for the winter, then harden off and plant out in their final positions in spring.

Cuttings are the usual method for the non-flowering chamomile used in lawns, but any of the species can also be treated this way. Use heel cuttings about 8 cm (3 in) long, taken in mid or late spring. Root under a cold frame, pot up singly when rooted (after about three or four weeks), then treat as seedlings.

Division is the easiest and most convenient method of propagation. This can be done in autumn or spring.

SOME POPULAR SPECIES	
A. cupaniana (Italy) Spreading mound of finely dissected, aromatic, grey leaves. White flowers, 5 cm (2 in) across, on erect stems, all summer. 30 cm (1 ft). **A. nobilis** *(Chamaemelum nobile)* (Europe) (Common chamomile) Mat-forming plant with aromatic, finely dissected green leaves. Daisy-like flowers about 4 cm (1½ in) across. The variety 'Treneague' does not flower and this is the form usually used for lawns. 25 cm (10 in).	**A. sancti-johannis** (Bulgaria) Lobed, fern-like grey-green leaves. Bright orange disc-like flowers in mid and late summer. 60 cm (2 ft). **A. tinctoria** (Europe) (Ox-eye chamomile) Deeply lobed and toothed green leaves with flowers about 5 cm (2 in) across, all summer. The species is seldom grown as the varieties are better. These include 'Wargrave Variety' (creamy-yellow), and 'Mrs E.C. Buxton' (lemon-yellow). 75 cm (2½ ft).

Top *Anthemis cupaniana*
Bottom *A. tinctoria* 'Grallagh Gold'

Aquilegia
Columbine, granny's bonnet
H/O/FS–PS

A genus of about 100 species. Some of them are small and suitable for the rock garden, but there are a couple of species that have a place in the herbaceous border, the hybrids *A. vulgaris* being traditional border plants. The name comes from the Greek *aquila* (eagle), as the spurred flowers can resemble an eagle's claws.

The plants are not long-lived, but are easily raised from seed.

How to grow
Plant in bold clumps, preferably in full sun, although partial shade is well tolerated. They are not fastidious about soil, but are best in fairly heavy loam or a moisture-retaining soil, provided it does not become waterlogged in winter. Incorporate plenty of humus.

Mulch in spring. Staking is not necessary. Dead-head individual blooms if possible, and cut the flower stems to ground level when they have finished.

Do not lift and replant more often than necessary as they resent disturbance and are best left to form big, strong clumps.

Propagation
Seed is the main method of propagation as the roots do not divide well. Surface sow in containers in mid or late summer. Place in a cold

One of the many attractive hybrids of Aquilegia

frame to germinate (which will take one or two months). Prick into individual pots and over-winter under the frame. Pot on into 10 cm (4 in) containers in mid spring, harden off in late spring, and plunge outdoors to grow on until planting time in autumn.

Alternatively, sow in warmth (18–21°C/65–70 °F) in early spring.

Root division is less satisfactory, but can be tried in mid autumn or early spring.

SOME POPULAR SPECIES	
A. caerulea (USA) Pale green almost fern-like leaves. Pale blue and white flowers with long spurs, mid spring to mid summer. 'Crimson Star' has red flowers with white centres. 45–60 cm (1½–2 ft). **A. vulgaris** (hybrids) Grey-green leaves resembling those on a	maidenhair fern in shape. The species is seldom grown as the hybrids are much superior. The long-spurred flowers are in shades of crimson, blue, violet, yellow, and white, in late spring and early summer. There are several good seed-raised strains. 60–75 cm (2–2½ ft).

Artemisia
Wormwood
H/O(D, most)/FS–PS

A large genus of about 400 species, including hardy and half-hardy herbaceous perennials and evergreen and semi-evergreen shrubs and sub-shrubs. The genus was named after the Greek goddess Artemis. Many species used to be used in medicine and cookery. But there are some

Artemisia ludoviciana

good garden plants, many with silvery, handsomely-cut foliage, although those described here are grown mainly for their flowers. Unfortunately some species may encroach on their neighbours.

How to grow

Most artemisias thrive on hot, dry sites, in poor, light, dry soil. *A. lactiflora* requires a moisture-retentive soil, but avoid wet or waterlogged ground for all of them. They also tolerate partial shade, but are best in full sun. Cut the plants almost to ground level in mid autumn.

Propagation

Heel cuttings, 5–8 cm (2–3 in) long, will root readily under a cold frame in mid and late summer. Pot up into individual pots before plunging outdoors until planting time in autumn.

Division in autumn or spring is the best and easiest method for herbaceous artemisias.

SOME POPULAR SPECIES	
A. lactiflora (China, India) (White mugwort) Leafy stems with deeply cut green pinnate leaves. Topped with plumes of fragrant creamy-white flowers in late summer and early autumn. 1.2 m (4 ft). **A. ludoviciana** (North America) (White sage) Woolly, white leaves. Erect habit. Silvery-white flowers in early and mid autumn, but really a foliage plant. May need curbing once established. 60–90 cm (2–3 ft). **A. nutans** (Sicily) Silvery-grey, divided,	evergreen leaves. Pale yellow flowers in late summer and early autumn, but these are not a feature. Spreads easily. 'Silver Queen', with willow-like silvery foliage, is a superior form. 60–75 cm (2–2½ ft). **A. stelleriana** (North America) (Dusty miller, old woman) Deeply lobed, aromatic, silvery leaves. Panicles of yellow flowers in late summer and early autumn. It can be rather rampant and is best lifted regularly and replanted. 60 cm (2 ft).

Asperula
H/O(M)/FS–PS

There are about 200 species, some of them hardy annuals. None of the species listed here is large, and *A. lilaciflora* will be happy in a rock garden. *A. odorata* is a rather wild plant that lacks the qualities of a good plant for the herbaceous border. The generic name comes from the Greek word *asper* (rough); the leaves feel rough.

How to grow

Although these plants like fairly moist conditions, avoid heavy or waterlogged ground. Otherwise they should be trouble-free, though *A. odorata* may have to be tidied regularly.

Propagation

Seed of *A. odorata* can be sown in early spring, in containers in a cold frame. Prick off into pots when large enough, singly or in pairs. Harden off before plunging outdoors in late spring or early summer. Plant out in early autumn.

Division is the simplest method of propagation for all of them. This is normally done in early or mid autumn.

Asperula lilaciflora caespitosa

SOME POPULAR SPECIES	
A. hexaphylla (Italy) Whorls of lanceolate leaves. Small pink flowers in early and mid summer. 23–30 cm (9–12 in). **A. lilaciflora caespitosa** (Eastern Mediterranean) A carpeting plant forming a mat of dainty, bright green foliage. Deep pink flowers in early and mid summer. 5–8 cm (2–3 in).	**A. odorata** *(Galium odoratum)* (Europe, Siberia) This plant should strictly be listed as a galium, but the old name of asperula still prevails. Whorls of linear leaves. Small fragrant white flowers in late spring and early summer. 15–23 cm (6–9 in).

Asphodeline
H/O/FS

King's spear, yellow asphodel. The genus derives its name from the fact that the plants are similar to the asphodelus. These are tall and stately plants providing a conspicuous display and best in a prominent position.

How to grow
Not demanding about soil, but requires a position in full sun.

Propagation
Seed can be sown under a cold frame in early spring, but germination is generally slow and erratic. Sow in containers about 8–10 cm (3–4 in) deep, and pot off into 13 cm (5 in) pots. Give the seedlings winter protection; they will be ready for planting in about another two years.

Asphodeline lutea

Division in autumn or spring is much easier. Lift carefully and take care not to pierce or lacerate the thick, fleshy roots.

SOME POPULAR SPECIES	
A. lutea *(A. luteus)* (Sicily) Strap-like narrow, drooping, rather grassy leaves. Stiff spikes of	fragrant, sulphur-yellow flowers produced all summer. 90–120 cm (3–4 ft).

Aster

H/O/FS

The name comes from the Greek *aster* (star), and reflects the shape of the daisy-like rayed flowers. There are about 500 species of aster, but the Michaelmas daisies *(A. novae-angliae* and *A. novi-belgii)* dominate the border species. There are, however, others equally attractive.

Michaelmas daisies have a rather poor reputation, often being regarded as tall, lanky plants, but some varieties are very compact and require no staking.

How to grow
Few plants respond so well to a little extra attention. Do not overcrowd the plants, and once established thin out the shoots on all but the youngest plants. Divide after three years.

Provide fertile soil and feed during the summer. Water freely if the ground is dry during the growing season. Tall varieties will need staking and this should be done early.

Aster 'Sunset'

Propagation
Seed is little used—the results are variable and the progeny often inferior.

Cuttings of new growth can be taken in spring. Make these about 8 cm (3 in) long, and root in containers under a cold frame. When rooted (usually in about three or four weeks), pot up singly and plunge outdoors to grow on. Plant out finally in autumn.

Division is the easiest and normal method of propagation. This is best done in mid or late autumn or early or mid spring. Ruthlessly discard the central worn-out portion of each clump.

<table>
<tr><td colspan="4" align="center">SOME POPULAR SPECIES</td></tr>
</table>

A. x amellus (Italy)
Grey-green, rather rough, leaves. Flowers about 5 cm (2 in) across in late summer and early autumn. There are varieties in various shades of blue, pink, and lavender. 60 cm (2 ft).

A. x frikartii (Garden origin)
A hybrid between the previous species and *A. thomsonii*. 5 cm (2 in) lavender-blue flowers,

mid summer to early autumn. 90 cm (3 ft).

A. novae-angliae (North America)
(Michaelmas daisy) Dull green, lanceolate leaves. Heads of daisy-type flowers, mainly in shades of pink and lavender, up to 5 cm (2 in) across. Generally taller and stiffer than the next species, which is the one more generally called Michaelmas daisy. It is

only the varieties that are generally grown, not the species. Early and mid autumn. 1.2–1.5 m (4–5 ft).

A. novi-belgii (North America)
(Michaelmas daisy) Mid green lance-shaped to linear leaves that clasp the stem. There are many varieties, with heads of daisy-type flowers in shades of mauve, blue, purple, red, pink, and

creamy-white. There are double and semi-double forms. Tall varieties: 75–120 cm (2½–4 ft); dwarf varieties: 30–60 cm (1–2 ft).

A. thomsonii (Himalayas)
Neat, grey-green bushes with pale lavender-blue, yellow-centred flowers, mid summer to early autumn. The variety usually grown is 'Nana'. 45 cm (1½ ft).

Astilbe

H/O(M)/FS–PS

The name comes from two Greek words: *a* (no) and *stilbe* (brightness), some species having drab flowers. Do not let that put you off growing the species and hybrids listed here. They are superb, colourful, plants.

Astilbes are still sometimes confused with spiraeas. They are exceedingly good plants for damp soil and other naturally damp and partially shaded positions, and are effective poolside plants.

Astilbes are suitable for a moist spot

The *A.* x *arendsii* hybrids are usually sold under their varietal name alone (e.g. *Astilbe* 'Rhineland').

How to grow

A good, moist loam is necessary for best results. Water freely in dry weather if not in a naturally moist position. Mulch plants in spring. Once planted, leave undisturbed for several years.

Cut the plants to ground level in mid autumn and divide every three years.

Propagation

Seed is not a very practical method of propagation, but it can be fun. Sow in spring and germinate at 16–18 °C (60–65°F): Prick out into individual pots, and move to a cold frame for the summer. Pot on the following spring, harden off, then plunge outdoors until ready for planting.

Division is easier, and you can be sure of the variety. Cut any old foliage to within 5 cm (2 in) of the ground and divide the plants in early or mid spring, replanting well-rooted portions.

<table>
<tr><td colspan="2" align="center">SOME POPULAR SPECIES</td></tr>
</table>

A. x arendsii (Garden origin)
Deep green, fern-like leaves. Feathery pyramidal plumes of tiny flowers from early to late summer. There are many varieties

in shades of pink, red, and white. 60–90 cm (2–3 ft).

A. chinensis (China)
Fern-like mid green leaves. Stubby spikes of lilac-rose flowers, mid to late summer. 30 cm (1 ft).

Astrantia

Masterwort
H/O(M)/FS–PS

A genus of ten hardy herbaceous perennials. They are not showy plants and most have flowers with subdued colours. They are however good cut flowers, and the plants will often tolerate quite poor conditions. The name comes from the Greek *aster* (a star), referring to the flowers.

How to grow
Best in partial shade, but will grow in full sun if the soil is kept reasonably moist in summer.

Although support is seldom required, in exposed positions some twiggy sticks may be necessary to prevent plants from being flattened. Cut down the stems in late summer.

Propagation
Seed is an easy method. Sow in containers in early autumn, and place in a cold frame to germinate. Prick out the seedlings into individual pots in spring. Pot on in summer, and plunge outdoors to grow on. Give winter protection and plant out in spring.

Division is both quick and easy. Lift large clumps to divide in early autumn or in early or mid spring. Replant only young outer portions.

Astrantia major

SOME POPULAR SPECIES	
A. carniolica (Europe) Green, divided foliage, and sprays of greenish-white flowers tinged pink, in mid summer. 60 cm (2 ft). **A. major** (Europe) A vigorous plant with three-lobed leaves and flat	heads of curious pinkish-green shaggy flowers on leafy stems, all summer. 60 cm (2 ft). **A. maxima** (East Caucasus) Light pink flowers, larger than in the previous species. 60 cm (2 ft).

Aubrieta

Purple rock cress
H/AK(D)/FS

A genus of 15 species of low-growing hardy ever-green perennials, commemorating M. Claude Aubriet, a French botanical artist. Although usually grown as rock plants, they are also excellent general carpeters—and make a fine informal edging for borders.

How to grow
Most soils are tolerated, but best results are likely to be on limy ground.

Cut the plants almost to the base after flowering to keep them neat.

Aubrieta, one of the most popular rock plants

Propagation

Seeds provide an easy method of propagation for a large number of plants. Sow in containers in warmth, in late winter and early spring. Prick out into individual pots. Harden off in late spring, then plunge outdoors for the summer. Plant out in flowering positions in early autumn.

Cuttings are more tricky. Take 5 cm (2 in) cuttings from the young growths in early or late summer. Root in a cold frame, and pot up singly in spring. Harden off before plunging outdoors in late spring to grow on.

Division is easy if the plants are prepared in advance. Trickle a 3–4 cm (1–1½ in) layer of fine potting compost between the stems of plants cut back after flowering. The plants will root into this and can be severed and lifted in late summer or early autumn.

SOME POPULAR SPECIES	
A. deltoidea (Sicily to Asia Minor) Spreading, mat-forming plant with masses of flowers in spring. Colours	include shades of red and purple, and there are named varieties. Some have double flowers. Spring. 8–10 cm (3–4 in).

Bergenia

H/O/FS–PS

Named in honour of Karl August von Bergen (1704–60), a German botanist, these are large-leaved, strong-growing plants, with practically evergreen leaves—assuming reddish winter tints in some cases.

They need no staking, are a good ground cover, make first-class tub plants, and are ideal front-of-border plants if planted in bold drifts.

Bergenia cordifolia

How to grow

Very easy to grow in most soils, in sun or partial shade. But they will respond to good cultivation. Feed the plants each spring and water copiously in dry weather. Leave the clumps undisturbed until they have to be lifted and divided because of overcrowding.

Propagation

Seed is rarely used, but if obtainable can be sown in warmth in mid winter. The young plants will be ready for planting out in a couple of years.

Division is the main method of increase. Lift large clumps in early or mid autumn, or after flowering in spring, and remove rooted offsets from the young outer edges.

SOME POPULAR SPECIES	
B. cordifolia (Siberia) Heart-shaped purplish fleshy leaves and large drooping sprays of pink flowers produced in mid and late spring. 30 cm (1 ft). **B. crassifolia** (Siberia) Ovate leaves, mahogany-tinted in winter. Panicles of lavender-pink bells from late winter to mid spring. **B. purpurascens** (Himalayas) Similar to previous species, but narrower	leaves. Purplish-pink flowers on purplish stems, mid and late spring. 30 cm (1 ft). **B. stracheyi** (Himalayas) Prostrate, rounded, hairy leaves, smaller than those of the other species listed, and less suitable for ground cover. The pink flowers passing to white at the margins, are borne in loose heads, in early and mid spring. There are named hybrids with a wider colour range. 23–30 cm (9–12 in).

Caltha

Kingcup, marsh marigold
H/AD(M)/FS–PS

A genus of 20 hardy herbaceous perennials, taking its name from the Greek *kalathos* (goblet) because of the shape of the flowers.

Really bog or waterside plants, and only successful in the herbaceous border if kept constantly moist.

The 'petals' are, in fact, petaloid sepals—but this is an academic point because they make lovely bright flowers in spring.

Caltha palustris 'Plena'

How to grow

A moist spot is really essential. Avoid a very hot, dry position. A neutral or acid soil is best, and adding plenty of moist peat at planting time will help.

Mulch annually with moist peat. If the leaves become unsightly as the season progresses, cut them off.

Propagation

Seed is best sown in summer in pots of soil-based seed compost placed in a cold frame with half the depth of the pot in clean water. Prick out the seedlings into individual pots, and again stand in shallow water. Pot on in spring, and grow on as before until early autumn, when they should be ready for planting out.

Division is easy and the normal method of propagation for *C. palustris.* In late spring or early summer, once flowering is over, lift large plants and split up the crown into several pieces.

Runners provide a useful means of increase for *C. polypetala.* Sever rooted young plantlets from the parent crown in summer.

SOME POPULAR SPECIES	
C. palustris (Europe) Deep green, rounded leaves. Goblet-shaped yellow flowers, mid to late spring. 'Plena' has button-like double flowers. 'Alba' is white. 30 cm (1 ft).	**C. polypetala** (Caucasus, Turkey) A plant with lush growth and sprawling stems. Yellow saucer-shaped flowers in mid and late spring. 60 cm (2 ft).

Camassia

Quamash

B/O(M)/FS

A genus of five bulbous plants more likely to be found in bulb catalogues than a nurseryman's list, but worth searching out. Large pyramidal spikes of blue flowers in mid summer. Both generic and common name are derived from *Quamash*, a North American word used for an edible species.

Camassia leichtlinii

How to grow

Best in moist, heavy soil. Water in spring and early summer during dry periods.

Plant the bulbs about 10 cm (4 in) deep in autumn, with at least 15 cm (6 in) between the bulbs. There is no need to lift the bulbs each year, but let the leaves die down naturally before tidying the plants. Remove any seed pods.

Propagation

Seed is a slow method, the plants taking three to five years to flower. Sow in pots under a cold frame in summer. It is important to use fresh seed. Pot up seedlings into individual 13 cm (5 in) pots when ready, usually about 15 months after sowing. Plunge the pots outdoors, but give winter protection. Plant out in autumn.

Offsets are easier and quicker, but still take two or three years to reach flowering size. Lift mature clumps in early autumn and carefully remove the offsets.

SOME POPULAR SPECIES	
C. cusickii (North Western USA) Bluebell-like leaves. Tall spikes of starry flowers on stiff, upright stems, early and mid summer. 90 cm (3 ft). **C. leichtlinii** (Oregon) Similar to previous species. Erect habit with blue or white flowers, early	and mid summer. 90 cm (3 ft). **C. quamash** (California, Canada) (*C. esculenta*) Bluebell-type foliage. Spikes of blue, purple, or white flowers, carried well clear of the foliage in summer. The bulb is edible. 75 cm (2½ ft).

Campanula

H/O/FS–PS

A large genus of about 300 species, many of them rock garden plants, some tender pot plants, and a few excellent border plants. The name comes from the Latin *campanula* (little bell).

How to grow

Mostly very easy plants to grow, thriving in any well drained soil in full sun or partial shade. But avoid an exposed position for the tall species, otherwise wind may be a problem. Support from twiggy sticks is necessary for most of them.

Dead-heading is worthwhile.

Propagation

Seed is fairly successful. Sow in containers in a cold frame in late spring or early summer and prick the seedlings out individually into small pots. Overwinter under the frame, and harden off before planting out in mid spring.

Cuttings root readily in a cold frame in mid spring. Make them 4–5 cm (1½–2 in) long, using basal shoots. Pot up individually, and pot on about a month later then plunge outdoors.

Division of most perennial species is successful and can be done in autumn.

Campanula latifolia 'Alba'

SOME POPULAR SPECIES	
C. glomerata (Europe) Lower leaves egg-shaped. Dense clusters of purple or blue bell-like flowers all summer. There is also a white variety. 30–60 cm (1–2 ft). **C. lactiflora** (Caucasus) Heads of lavender-blue, pink, or white flowers, mid to late summer. 90–120 cm (3–4 ft). There is a dwarf variety—'Pouffe'— 30 cm (1 ft). **C. latifolia** (Europe) (Giant bellflower) Spikes	of tubular, purple-blue flowers like narrow bells, all summer. Other shades of blue as well as white. 1.2 m (4 ft). **C. persicifolia** (Europe, Siberia) (Peach-leaved bellflower) Clumps of evergreen leaves forming rosettes, topped with clusters of saucer-shaped flowers on wiry stems, most of summer, blue and purple-blue, as well as white. 60–90 cm (2–3 ft).

Catananche
Cupid's dart
H/O(D)/FS

The common name for this plant is a reflection of the Latin name, which is based on the Greek *katananke* (strong incentive), an indication of its use in love potions. It is a small genus of five annuals and herbaceous perennials, but only one species is in general cultivation.

The plant is usually short-lived, but is easily raised from seed. The flowers are good for cutting and can be dried for winter decoration.

How to grow
Best on dry, sandy soils, and able to withstand drought well. It will need the support of twiggy sticks if planted in an exposed position.

Propagation
Seed is the best method of propagation. Sow in warmth (13–16°C/55–60°F) in later winter or early spring. Prick out singly into small pots, harden off, and plant out in late spring where they are to flower.

Alternatively, sow in containers under a cold frame in mid or late spring. Prick out into small pots, overwinter in the frame, and plant out the following spring after hardening off.

Root cuttings are quick to root under a cold frame if taken in late winter or early spring. Make them about 5 cm (2 in) long, and pot up

Catananche caerulea

individually into 10 cm (4 in) pots when new leaves emerge (usually after six to eight weeks). Plunge under the frame, and grow on. Leave the frame lights (tops) off during the summer, and plant in their flowering positions in autumn.

SOME POPULAR SPECIES	
C. caerulea (Western Mediterranean) Greyish, grassy leaves. Silvery buds open to blue papery flowers about 4 cm (1½ in) across supported	on individual long wiry stems, mid summer to early autumn. 'Major' has larger, lavender flowers. There are other varieties. 60 cm (2 ft).

Centaurea
H/O/FS

A large genus of about 600 species, deriving its name from classical Greek myths. The plant is said to have healed the wound of Chiron, one of the Centaurs. It contains popular hardy annuals such as cornflowers and sweet sultans, tender species, and of course some very useful border plants.

The flowers are generally good for cutting.

Centaurea dealbata 'Steenbergii'

How to grow

Centaureas will put on a respectable show on poor, hungry soil, but will do much better on fertile land. They respond to a little extra attention.

Tall species may need the support of twiggy sticks. Dead-head regularly to prolong the display.

Divide after three years as the plants spread rapidly.

Propagation

Seed can be sown in containers in a cold frame in mid spring. The seedlings should be ready to prick out into small pots after about a month. Harden off before plunging outdoors in early summer. Give winter protection, then harden off and plant out in their flowering positions in mid or late spring.

Division in mid autumn or early spring is easier and quicker, especially if only a few plants are needed.

SOME POPULAR SPECIES	
C. dealbata (Caucasus, Persia) Jagged silvery-grey leaves, topped with fringed flowers about 6 cm (2½ in) across in early and mid summer. The varieties usually grown are 'John Coutts' (rose-pink) and 'Steenbergii' (deep pink). 60–75 cm (2–2½ ft). **C. macrocephala** (Armenia, Caucasus) A rather stiff-looking	plant with rough light green leaves and yellow thistle-like flowers on thick, leafy stems in early and mid summer. 1.2–1.5 m (4–5 ft). **C. montana** (Pyrenees, Alps, Carpathians) Silvery foliage and feathery blue flowers, like giant cornflowers. Late spring to mid summer. There are white, pink, and purple varieties. 45 cm (1½ ft).

Centranthus

Valerian

H/AK/FS

A genus of 12 species of annuals and hardy herbaceous perennials, although only one species is in general cultivation. The genus takes its name from the Greek *kentron* (a spur) and *anthos* (a flower), alluding to the flower shape.

Although short-lived, the species described is easily raised from seed.

How to grow

Among the easiest plants to grow. They will thrive almost anywhere—self-seeded plants often establish themselves on dry walls. And although most soils are tolerated, this plant is especially useful for anyone gardening on very chalky ground.

Cut down the stems to ground level in autumn.

Propagation

Seed is the usual method of propagation. Sow in containers in a cold frame in mid or late spring. Prick out the seedlings into small pots, harden off, and plunge outdoors. Give winter protection and set the plants out in their flowering positions in mid spring.

SOME POPULAR SPECIES	
C. ruber (Europe) Shiny, lance-shaped leaves. Bold panicles of small pink flowers carried well clear of the foliage, all	summer. There are varieties with crimson or white flowers. 45–75 cm (1½–2½ ft).

Centranthus ruber

Cerastium

Snow in summer
H/O(D)/FS

A genus of about 60 species of free-flowering hardy annuals and herbaceous perennials. Most are weedy and invasive, but those listed here can be very useful in the right situation. They are very easy plants to grow, but too rampant to put close to small plants that may be smothered. They are, however, very useful for providing ground cover at the front of the border or for clothing dry banks or rock beds.

The name comes from the Greek *keras* (horn), alluding to the horn-shaped seed capsules.

How to grow
A sunny position and well-drained soil suit them best, but cerastium will grow reasonably well on most soils.

Cut the plants back as necessary to keep them within bounds before other plants are encroached.

Propagation
Seed is an easy method of propagation. If sown in early or mid spring in warmth (18–21°C/ 65–70°F), the seedlings should show in two or three weeks. Prick out individually into small pots, and harden off before plunging outdoors in early summer. Plant in flowering positions in early or mid autumn.

Alternatively, sow the seed in a cold frame in early spring. Plunge outdoors for the summer, but overwinter under the frame and finally plant out in mid spring.

Division of large clumps in mid or late spring is possible. Carefully lift a mature plant and gently tease away rooted portions from around the edge.

SOME POPULAR SPECIES	
C. biebersteinii (Crimea) Carpeting plant with small, woolly grey leaves, and white flowers in late spring and early summer. 10–15 cm (4–6 in).	**C. tomentosum** (Southern and Eastern Europe) Similar to previous species but leaves smaller and less woolly. 10 cm (4 in).

Cerastium tomentosum, snow in summer, makes a superb but invasive ground cover

Chelone

Turtle-head
H/O(M)/FS–PS

A small genus of four hardy herbaceous perennials, the name being derived from the Greek *kelone* (a tortoise), alluding to the shape of the helmet of the flower. This is similar in outline to the head of a turtle or tortoise.

Chelones are useful plants for a mixed border as well as the traditional herbaceous border. There is likely to be confusion in the naming of the species listed here within the trade, and the plant sometimes offered as *C. lyonii* may be a smaller plant than suggested here.

How to grow
Best in a sunny position on light soil, although good results can be expected from plants grown in partial shade.

The plants may need staking in an exposed position. Cut the dead stems down in mid autumn.

Propagation
Seed is not the easiest method, but can be sown in warm (16°C/60°F) in early spring. Prick out into small pots and harden off before plunging outdoors in late spring or early summer. Provide winter protection, pot on in mid spring, and harden off before plunging outdoors again in late spring. Plant in flowering positions in autumn.

Cuttings root easily in a cold frame if taken in early or mid summer. Choose new non-flowering shoots and make the cuttings about 8 cm (3 in) long. They should root in about a month, then pot up singly into small pots. Overwinter under a frame and treat as seed-raised plants. Divide large plants in autumn or spring.

SOME POPULAR SPECIES	
C. lyonii (North America) Dark green leaves in opposite pairs. Rose-purple snapdragon-like flowers in terminal clusters all summer. 90 cm (3 ft).	**C. obliqua** (North America) Bright green leaves with prominent veins. Deep rose flowers on stiff erect stems. Mid summer to early autumn. 60 cm (2 ft).

Chelone obliqua, with its snapdragon-like flowers, is one of the more unusual border plants

Chrysanthemum
H/O/FS

A genus of more than 200 species, just a few of which are used as border plants. Among the late-flowering ones in particular, it is mainly hybrids that are grown. The name comes from the Greek *chrysos* (gold) and *anthemon* (flower), but among the late-flowering chrysanthemums pinks and reds are just as dominant.

This diverse group of plants can be confusing if you do not know the various types. The Shasta daisy *(C. maximum)* is summer-flowering and a popular border plant with its large daisy-type flowers.

The autumn-flowering types are the *C. rubellum* varieties (single flowers) and the Korean and border chrysanthemums. Only those chrysanthemums used for general border decoration have been included here, although they are, of course, suitable for cutting.

How to grow

Chrysanthemums prefer a well-drained, fertile soil in a sunny position, and do especially well on chalky soil. *C. rubellum* may need staking.

Cut the stems back to ground level in early winter.

None of the varieties grown as border plants should be disbudded—a mass of small flowers will be far better than fewer but larger blooms for a garden display. In the border, plant in blocks of three or five of the same variety. When the plants are 15–25 cm (6–10 in) tall, pinch out the growing tip to encourage bushiness (this may not be necessary with some spray varieties as they branch out naturally).

Although late-flowering types mentioned here are almost hardy, and should be all right left in the ground if covered with a layer of peat or pulverised bark for the winter (add a few slug pellets too), they may succumb in damp, wet soil. It is best to lift the plants about a month after flowering has finished, and overwinter the stools in a cold frame or cold greenhouse. Do not let the soil dry out completely, but avoid too much moisture.

Use the stools to provide a supply of cuttings in spring.

Chrysanthemum rubellum

Propagation

Seed can be used, but results are likely to be very variable; cuttings or division will give more predictable results.

Cuttings of 5–8 cm (2–3 in) long basal non-flowering shoots will root readily in a cold frame in early spring. To ensure plenty of early cuttings, pot up and overwinter the stools under cover. Root the cuttings around the edges of small pots, then when rooted pot up singly. Harden off and plant out in late spring.

Division in early or mid spring is the best method for *C. maximum*.

SOME POPULAR SPECIES	
C. maximum (Pyrenees) (Shasta daisy) Large, daisy-like flowers about 8–10 cm (3–4 in) across with a prominent yellow eye, blooming most of the summer. There are semi-double and double varieties. 60–90 cm (2–3 ft). **C. rubellum** (Japan) Single flowers about 5 cm (2 in) across, from late summer to mid autumn. Colours include pink,	gold, yellow, and crimson. 60 cm (2 ft). **Border chrysanthemums** There are several types of late-flowering chrysanthemum that make good border plants—the Korean hybrids and outdoor spray varieties in particular. It is best to consult a specialist chrysanthemum catalogue for details of these.

Cimicifuga

Bugbane

H/O(M)/FS–PS

A genus of 15 species of hardy herbaceous perennials, which despite the ugly common name include some beautiful plants. The name is derived from the Latin *cimex* (bug) and *fugio* (to run away): one species was used in Russia to drive away bed bugs.

How to grow

Best in partial shade, and moist soil containing plenty of humus. Mulch with leafmould or pulverised bark each spring.

Will need support in an exposed position. Cut down the stems in late autumn.

Propagation

Seed is little used but is obtainable. Sow in containers under a cold frame in early or mid spring. Prick out into small pots when large enough (it may take one to three months to reach this size). Give winter protection, then pot on in mid spring. Harden off and plunge outdoors in late spring or early summer. Plant in final positions in autumn, but in cold areas protect for another winter then plant out in spring.

Division is quicker and easier than seed. Lift and split up large clumps in early or mid autumn, or wait until early spring.

SOME POPULAR SPECIES	
C. americana *(C. cordifolia)* (North America) Some botanists put *C. cordifolia* as a variety of *C. racemosa*, but horticulturally the plant is likely to be found under *C. americana.* Divided, almost fern-like leaves. Slender spikes of small creamy-white flowers in late summer and early autumn. 60–120 cm (2–4 ft). **C. dahurica** (Japan) Slender spikes of tiny ivory flowers, late summer and	early autumn. 1.2–1.5 m (4–5 ft). **C. japonica** (Japan) Deeply cut, wide-spreading foliage. Thick white pokers, late summer and autumn. 1.2 m (4 ft). **C. racemosa** (North America) (Black snake root) Creamy-white plumes on slender arching stems, late summer to mid autumn. 1.2 m (4 ft). **C. ramosa** (Japan) Tall tapering creamy-white pokers in early autumn. 1.8 m (6 ft).

The unusual bottle brush-like flowers of Cimicifuga racemosa

Colchicum

Autumn crocus
B/O/FS–PS

Colchicum speciosum

The autumn crocuses are an unusual group of plants—the flowers appear before the foliage—and they are not really crocuses at all despite their appearance. The genus takes its name from a province in Asia Minor called *Colchis* (from where some species originate).

Place them near the front of the border where they will be appreciated when their large crocus-shaped flowers appear in autumn when most other plants are finishing. They are especially good among deciduous shrubs and planted in grass (grass has the advantage of keeping the blooms cleaner if it rains—they tend to become flattened and splashed with mud in heavy rain).

How to grow
Easy to grow, but they respond to a rich soil. Plant as early as possible—in mid summer if you can—covering the bulbs with at least 10 cm (4 in) of soil.

Propagation
Seed is slow, but necessary if you want a large number of plants. Sow in pots of soil-based potting compost in a cold frame in early or mid summer. Germination is usually erratic and may take 18 months. Leave the seedlings in the pots to grow on for about a year, then plant out in a nursery bed in mid or late summer. Alternatively,

pot up three seedlings to a 20 cm (8 in) pot and subsequently repot each summer. But bear in mind that small corms may take four to six years to reach flowering size.

Offsets (small corms) are an easier method. Remove them from mature corms about mid summer, after the leaves have died down. Plant out in nursery beds, or pot up, to grow on. They should flower in another four years.

SOME POPULAR SPECIES	
C. agrippinum (Origin uncertain) Flowers mottled reddish-purple on a whitish background, early autumn. Semi-prostrate leaves appear in spring. 10–15 cm (4–6 in).	**C. byzantinum** (Asia Minor) Large lilac-pink flowers, early autumn. 15 cm (6 in), the broad leaves doubling this height when they appear.
C. autumnale (Europe) Lilac flowers, but there are also white and pink varieties, some double, early and mid autumn. 15 cm (6 in), leaves a couple of inches longer when they appear.	**C. speciosum** (Persia, Caucasus, Asia Minor, Syria, Lebanon) Large, erect, goblet-shaped flowers in early and mid autumn. 15 cm (6 in), the leaves doubling this height when they appear.

Convallaria

Lily of the valley
H/O/PS–SD

A genus with just the one species, taking its name from the Latin *convallis* (valley). Popular for its fragrant flowers, which also cut well, the lily of the valley is unfortunately not an ideal border plant—it is best in a wild garden or in a cool, shady position.

Convallaria majalis

How to grow

Once established, lily of the valley can produce a spreading carpet of fragrant flowers, but often they fail to thrive. Although partial shade or shade is best, they will not compete with tree roots.

They do best in heavy soil enriched with well-rotted manure or compost, or perhaps moist peat.

Plant in early autumn, spreading the stoloniferous roots out about 8–10 cm (3–4 in) below the surface. Make sure the soil is moist from spring to autumn, and mulch annually.

Propagation

Seed, if obtainable, should be sown in early autumn in 15 cm (6 in) half-pots of soil-based seed compost in a cold frame. Uncover the frame when the seedlings are 5 cm (2 in) tall, but leave them undisturbed in their containers until the foliage dies down in early autumn. Then plant out the young 'pips' in a nursery bed. Grow on for two or three years, by which time they should have reached flowering stage.

Division is quicker and easier. Lift and split up the clumps in either early spring before growth starts, or in early or mid autumn when the leaves have died down. Replant individual crowns about 10–15 cm (4–6 in) apart in groups in a well-prepared bed.

SOME POPULAR SPECIES	
C. majalis (Europe, Asia, North America) Pairs of lance-shaped leaves emerge from the crown, embracing arching flower stems with very fragrant small white bells	in late spring. There is a variety with larger flowers—'Fortin's Giant'—and one with variegated foliage. 15–23 cm (6–9 in).

Coreopsis

H/O(D)/FS

A genus of more than 100 species, but only a few are worthwhile perennial border plants. The name is derived from the Greek *koris* (a bug or tick) and *opsis* (like), alluding to the bug-like appearance of the seeds.

The plants listed here are best in bold groups of one variety. They tend to be short-lived, but are not difficult to propagate.

The flowers are good for cutting.

How to grow

Coreopsis do best on well-drained, fertile soil in full sun. They will probably need the support of twiggy sticks from an early stage.

Cut down to ground level in autumn.

Propagation

Seed can be sown in containers under a cold frame in mid spring. Seedlings are usually ready to prick out singly into small pots in about six weeks. Harden off before plunging outdoors in mid summer. Plant out in mid autumn.

Alternatively sow in warmth (16–21 °C/ 60–70 °F) in late winter to flower the same year. Prick out seedlings individually into pots, and grow on a few degrees cooler. Harden off before planting outdoors in late spring.

Division is a popular and easy method of increase. Cut down the old stems to within 5 cm (2 in) of soil level in mid autumn or early spring, lift a mature clump, and tease off rooted pieces of stem from the outer edges. Discard the worn-out central portion.

SOME POPULAR SPECIES	
C. grandiflora (USA) Lance-shaped leaves above which arise a long succession of yellow rayed flowers for most of summer. There is a double variety, and a dwarf ('Goldfink') only 25 cm	(10 in) high. 45–75 cm (1½–2½ ft). **C. verticillata** (USA) Finely divided leaves on neat, bushy plants covered with starry, rich yellow flowers in mid and late summer. 60 cm (2 ft).

Coreopsis verticillata

Crambe

H/AK/FS–PS

A member of the cabbage family (*krambe* is the Greek name for cabbage). Naturally coastal plants, the species listed here is useful for seaside gardens. But it will do well in inland gardens too if given the right conditions.

Crambe cordifolia

C. cordifolia is a huge plant with large sprays of small white flowers best seen against a darker background of trees and shrubs.

How to grow
Easily grown on neutral or alkaline soil, but allow plenty of space.

Propagation
Seed is worth trying if you are prepared to wait— the seedlings may take three years to reach flowering size. Sow under a cold frame in early spring. Prick out into small pots, pot on in early summer, and plunge outdoors until ready to plant in autumn.

Root cuttings taken in mid autumn or late winter provide the easiest and most reliable method of increase. Make them 8 cm (3 in) long, inserting about half a dozen cuttings in a 13 cm (5 in) pot of sandy cutting compost. Pot up singly when rooted, harden off and pot on into 15 cm (6 in) pots or grow on in a nursery bed until autumn.

Basal cuttings and division in early spring are both possibilities but less practical than root cuttings.

SOME POPULAR SPECIES	
C. cordifolia (Caucasus) Basal large, rhubarb-like leaves. Broad yet graceful panicles of white flowers	and branched stems produced in early and mid summer. 1.5–1.8 m (5–8 ft).

Crinum

Cape lily
B/O/FS

A genus of more than 100 bulbous plants, but only a few are hardy enough to be attempted outdoors. The name is based on the Greek *krinon* (a lily).

How to grow
A deep, fertile but well-drained soil is required for these bulbs. Plant in spring in a warm, sunny spot, perhaps close to a sheltered wall, and cover

Crinum x *powellii*

with at least 15 cm (6 in) of soil. The emerging shoots may need protection from frost, perhaps with coarse sand or dry peat.

Water freely in summer.

Propagation

Seed is a challenge and the seedlings may take four or five years to reach flowering size. Offsets are a better proposition.

Lift mature bulbs in early spring (no more than once every three or four years), and carefully remove the offsets. Replant immediately.

SOME POPULAR SPECIES	
C. bulbispermum *(C. longifolium)* (South Africa) Strap-shaped spreading leaves. Large, wide-mouthed white flowers, sometimes flushed pale pink, produced in early and mid autumn. 45 cm (1½ ft).	**C. x powellii** (Garden origin) Long, strap-like leaves. Large lily-like flowers which open in succession over several weeks, from mid summer to early autumn. Pink is the usual colour, but white ones are available. 90 cm (3 ft).

Crocosmia

Montbretia

B/O(M)/FS–PS

An attractive group of border plants closely related to montbretia, with which they are often confused. *C.* x *crocosmiiflora* varieties are popularly known as montbretias. All make good cut flowers.

How to grow

Plant the corms in light sandy soil, enriched with well-rotted manure or compost. They are best planted in early autumn, although spring planting is another option.

C. masonorum is hardy, *C.* x *crocosmiiflora* nearly so (it will survive most winters in Britain), but as a safeguard it may be worth lifting some corms in autumn for replanting in spring. Clean the corms, removing the leaf remains when these have turned brown, and store in a frost-free place.

Propagation

Seeds are best sown, several to a 13 cm (5 in) pot, as soon as they ripen in early autumn. If placed in a cold frame they should germinate by spring. Do not disturb the seedlings, but let them grow on under the frame. Harden off and uncover about mid summer. When the foliage dies down in autumn, plant out the young corms about 8–10 cm (3–4 in) apart. They should flower in one or two years. In cold districts overwinter them under frames and plant out in the garden the following spring.

Crocosmia masonorum

Offsets are the easiest and quickest method of propagation. Lift and divide mature clumps every three or four years, in mid autumn or early spring.

SOME POPULAR SPECIES	
C. x crocosmiiflora *(Montbretia crocosmiiflora)* (Garden origin) Trumpet-shaped flowers clustered along wiry stems, in shades of yellow, orange and red, from mid summer to early autumn. 60–90 cm (2–3 ft).	**C. masonorum** (South Africa) Strap-like ribbed foliage. Trumpet-shaped orange flowers closely packed on arching wiry stems, in mid and late summer. 60 cm (2 ft).

Delphinium
H/O/FS

A genus of 350 species, including annuals as well as perennials. The main border delphiniums are hybrids derived from crosses between *D. elatum* and *D. grandiflorum*. These are split into two groups: Elatum varieties have an erect, stiff habit and large florets (sometimes double or semi-double), Belladonna varieties are smaller, more branching plants with wiry stems and cupped florets.

Delphiniums have a very keen following and there are books and societies dedicated to them. A brief account is all that is possible in the space available.

The name is derived from the Greek *delphin* (dolphin), presumably because the flower buds bear some resemblance to the overall shape of a dolphin.

A mixed group of delphinium hybrids

How to grow

Delphiniums demand rich soil and good drainage. Thin out weak stems from the centre of an established clump. Stake the rest of the shoots, angling the canes outwards slightly.

Slugs and delphiniums are not good companions, and slugs will make short work of these plants.

Mildew is an ever-present risk where delphiniums are grown. Watch out for it and spray with a suitable fungicide as soon as it is noticed.

After flowering, cut the spikes to the base—under favourable conditions they may flower again later, especially the Belladonna type.

Propagation

Seed is an easy method of propagation if you do not want named varieties (some Pacific hybrids do come true from seed), but it loses its vitality quickly so try to use fresh seed (store any left over in a dry atmosphere at about 2–4°C/36–40°F).

Sow in late winter in containers, barely covering with compost. Germination usually takes about 15–25 days at 13–16°C (55–65°F). Prick out the seedlings singly into small pots, and grow on a few degrees cooler. Harden off and plant out in their flowering positions in late spring.

Cuttings of new basal shoots will root readily in early and mid spring. Make them 8 cm (3 in) long, discarding any with hollow stems. They should have rooted and be ready for potting up after about a month. Harden off and plant out in late spring.

Division is simple and best done in early or mid spring.

SOME POPULAR SPECIES	
D. x hybridum (Garden origin) These hybrids, derived mainly from *D. elatum* and *D. grandiflorum* fall into three groups: *Elatum* These have tall, upright spikes of flat-faced flowers, sometimes double or semi-double. Heights vary from 1–2.4 m (3–8 ft). *Pacific Giants* These are	large and tall, and generally similar to the previous type. They are usually short-lived and are normally raised from seed each year. *Belladonna* These are smaller (1–1.2 m/3–4 ft), branching plants with more cup-shaped and widely spaced individual flowers.

Dianthus

Border carnations, pinks

H/AK/FS

A fairly large genus of about 300 species of annuals and perennials with evergreen leaves. The sweet William *(D. barbatus)* is a biennial not included here, but it is a useful gap-filler in the herbaceous border. Most of the species are more at home in the rock garden, and it is only the border carnations and modern hybrid pinks that you are likely to plant in the border.

The name dianthus comes from the Greek *dios* (divine) and *anthos* (flower), meaning divine flower.

How to grow

Although dianthus are associated with alkaline soils, they will flourish on neutral ground too. It is only acid conditions that they do not tolerate well. They are, however, a very good choice for chalky soils.

Propagation

Seed is an easy way to raise most dianthus. Sow in warmth in mid or late spring. They should germinate in 10–28 days at 18–21°C (65–70°F). Prick out into small pots and grow on under a cold frame for three weeks. Harden off and plunge outdoors, planting the seedlings out in their flowering positions in early autumn.

Cuttings also provide an easy method of propagation for pinks. Use non-flowering growths in early or mid summer, and make the cuttings 8–10 cm (3–4 in) long. Root in a sandy cuttings' compost in a cold frame. Pot up when rooted, and plant out in early autumn.

Layering is a popular way to increase border carnations, and the technique can also be used for pinks. Layer in early or mid summer, and allow about ten weeks before severing, lifting, and planting out.

SOME POPULAR SPECIES	
Border carnations (Garden origin) Narrow, grey-green leaves, and double flowers about 5 cm (2 in) across, often highly fragrant. They flower only once and the period is relatively brief—in mid or late summer. The plants are short-lived and should be replaced every couple of years. Wide colour range, many picoteed. 60 cm (2 ft).	**Modern pinks** (Garden origin) These originate from crosses between old-fashioned pinks and perpetual-flowering carnations. From the hybrid *D.* x *allwoodii* the modern pinks evolved. They flower in early and mid summer, and usually again in early autumn. They are short-lived and need to be replaced regularly. 30 cm (1 ft).

Dianthus 'Helen'

Dianthus 'London Poppet'

Dicentra

H/O/PS

A genus of 20 species of hardy herbaceous perennials, the name being derived from the Greek words *di* (two) and *kentron* (spur), alluding to two spurs on the petals.

The plants listed here are all useful front-of-border plants, their graceful dissected, fern-like foliage being a point of interest all summer, even when the plants are not in flower.

How to grow

A warm spot sheltered from strong winds is ideal. And although they need watering freely in dry weather, the ground should be well drained to avoid waterlogging in winter.

Dicentras will grow in any well-drained soil, but will do best if it is enriched with well-rotted manure or compost. Do not disturb the clumps more than necessary. Mulch each spring.

Propagation

Seed is fairly easy. Sow in warmth (16°C/60°F) in early spring, preferably after chilling in the fridge for six weeks beforehand. Prick out seedlings into small pots, harden off and plunge in a cold frame. Grow on until the following spring, when the plants can be set in their flowering positions.

Root cuttings taken in early spring root readily.

Dicentra spectabilis

Make them 8–10 cm (3–4 in) long, and root under a cold frame in slight warmth (10–13°C/50–55°F). Pot up singly when growth is evident (usually in four to seven weeks), harden off, and plunge outdoors for the summer. Plant in autumn. Division is the easiest method and can be done in autumn or spring.

SOME POPULAR SPECIES	
D. eximia (North America) Grey-green foliage. Rose-pink narrowly heart-shaped flowers in drooping racemes, mid spring to early summer. There is a white variety. 30 cm (1 ft). **D. formosa** (North America) Bright green feathery foliage. Narrowly heart-shaped pink flowers in	arching racemes in late spring and early summer. 30–45 cm (1–1½ ft). **D. spectabilis** (China, Japan) (Bleeding heart) Deeply incised feathery foliage. Graceful arching stems laden with rosy-pink and white pendent flowers resembling a heart-shaped locket, mid and late spring. 60–90 cm (2–3 ft).

Dictamnus

Burning bush

H/AK(D)/FS

A genus of six hardy herbaceous perennials, taking their name from *diktamnos*—a classical Greek name for the plant—perhaps because some were once common on Mount Dicte.

These are not spectacular plants, but quite attractive and the species in general cultivation has an interesting characteristic—it contains a volatile oil that on a warm, still summer's day

Dictamnus albus

(especially in the evening), can be ignited with a lighted match placed just beneath a flower spike. For a moment the spike will be enveloped in a leaping flame, but it quickly disappears, leaving the plant unharmed.

How to grow

This plant likes a sunny position, and does well on dry soils. Although it thrives on chalky land, you can expect good results on neutral soils too.

Root disturbance is resented, and the plant is best left undisturbed for as long as possible. It will take a season or two to make a respectable plant anyway.

Propagation

Seed is the normal method of propagation. Sow in containers under a cold frame or cloche in late summer or early autumn. Prick out the seedlings into small pots in spring (germination is erratic), and grow on under a frame. Pot on and harden off before plunging outdoors in late spring. Plant out about 16 months later (about two years from sowing).

Alternatively, chill the seeds in a fridge for four to six weeks before sowing in warmth (13–16°C/55–60°F) in mid or late winter. The seeds will probably take several months to germinate. Prick out and grow on as above.

SOME POPULAR SPECIES	
D. albus *(D. fraxinella)* (Southern Europe, Asia) Ash-like, deeply cut lemon-scented foliage,	topped with spikes of fragrant white or purple-pink flowers in mid and late summer. 60 cm (2 ft).

Digitalis
Foxglove
H/O/PS

A genus of about 30 species of hardy biennials and perennials. The popular foxglove *(D. purpurea)* will go on as a perennial sometimes, but it is best to sow a fresh supply each year, treating the plant as a biennial.

How to grow

Will do well on most soils if given partial shade, but try to make sure the ground does not dry out in summer.

Propagation

Seed is the accepted method of propagation. Sow on the surface in containers under a cold frame from mid spring to early summer. Prick out the seedlings singly into 10 cm (4 in) pots, harden off, then plunge outdoors until autumn, when they should be planted. These plants will flower the next year.

Some varieties will flower the first year if treated as half-hardy annuals. Surface sow in containers and germinate in warmth (16–18°C/60–65°F) in late winter. Pot up the seedlings into 10 cm (4 in) pots, harden off, and plant out in late spring.

SOME POPULAR SPECIES	
D. grandiflora *(D. ambigua)* (Europe, Caucasus, Siberia) Lanceolate leaves. Spikes of yellow flowers with brown netting, mid and late summer. A perennial species. 60–90 cm (2–3 ft). **D. x mertonensis** (Garden origin) Mid green, lanceolate leaves. Spikes of flowers the colour of crushed strawberries, all summer. Perennial. 75 cm (2½ ft).	**D. purpurea** (Western Europe) Oblong leaves forming a rosette. Tall, usually one-sided spikes of bell-like flowers. The modern varieties include shades of purple, red, maroon, pink, cream, and yellow, and are usually spotted, and the bells may be all round the spike. Biennial. 'Excelsior' is one of the best strains, but 'Foxy' can be treated as an annual. 1–1.5 m (3–5 ft).

Mixed colours from one of the Digitalis *hybrids*

Doronicum

Leopard's bane
H/O/FS–SD

A genus of 35 hardy herbaceous perennials. All those commonly grown have yellow, daisy-like flowers in early spring. These are long-stalked and useful for cutting.

How to grow
Very easy to grow, thriving in shade or sun, although heavy shade is best avoided. Almost any soil will do and they thrive on heavy ground. Plant the short rhizomes close to the surface.

It may be necessary to provide twiggy support for plants in an exposed position.

Regular dead-heading may produce a second flush of flowers in autumn. Cut down to ground level before winter. Feed annually, and control slugs.

Propagation
Seed is worth considering if you need a large number of plants. Sow in containers under a cold frame in early or late spring. Prick off into small pots, then harden off before plunging outdoors for the summer. Give winter protection, then plant out in mid spring.

Division is easy, and the usual method of propagation. This can be done after flowering in spring, or in early or mid autumn.

Doronicum plantagineum 'Harpur Crewe'

SOME POPULAR SPECIES	
D. austriacum (Central Europe) Bright green, heart-shaped leaves. Yellow daisy flowers about 6 cm (2½ in) across, early to late spring. 45–60 cm (1½–2 ft). **D. columnae** *(D. cordatum)* (Balkans, Asia Minor) Kidney-shaped bright green leaves that clasp the stem. Yellow daisy flowers	5 cm (2 in) across in early to late spring. 23 cm (9 in). **D. plantagineum** (South West Europe) Heart-shaped bright green leaves. Yellow flowers up to 8 cm (3 in) across, produced in mid spring to early summer. Good varieties are 'Miss Mason' (bright yellow), and 'Harpur Crewe' (golden yellow). 60–90 cm (2–3 ft).

Echinacea

Purple cone flower
H/O/FS

A small genus of three hardy herbaceous perennials, closely related to the rudbeckias. The name is derived from the Greek *echinos* (hedgehog), alluding to the whorl of prickly, pointed bracts beneath the flower head.

Echinaceas make good cut flowers, having long stems and good lasting qualities.

How to grow
Fertile, deep loam with good drainage is necessary for good results. Avoid ground prone to waterlogging.

Remove flower stems as the blooms fade to prolong the flowering season.

Echinacea 'Robert Bloom'

Propagation

Seed can be sown in warmth (13–16°C/55–60°F) in mid spring. Prick out the seedlings when they are ready—usually after about a month—and later move them to a cold frame. Pot on in mid summer, then plunge outdoors. Plant out in early autumn.

Root cuttings taken in late winter should present no problems. Make them 5–8 cm (2–3 in) long, inserting half a dozen in a 13 cm (5 in) pot. When the leaves show about mid spring, pot up singly and grow on under the frame. Pot on about mid summer and treat as for seed-raised plants. Division is quick and easy. Autumn or early spring are the best times.

SOME POPULAR SPECIES	
E. purpurea (North America) Rough, lanceolate leaves. Large rayed daisy-like reddish-purple flowers with a prominent central	cone. 'The King' is crimson-pink with a mahogany centre. There is also a white-petalled form. Mid summer to early autumn. 1–1.2 m (3–4 ft).

Echinops

Globe thistle

H/O(D)/FS

A genus of about 100 species of hardy herbaceous perennials and biennials. The name is taken from the Greek *echinos* (a hedgehog) and *opsis* (like), alluding to the spiky appearance of the flower heads, which look a little like rolled-up hedgehogs.

The spherical heads after the little corollas fall off resemble thistles, hence the common name.

The flowers are good for cutting, and can be dried.

How to grow

The basic requirements are a deep, rich soil with adequate drainage.

Leave the plants undisturbed for as long as possible unless root cuttings are required.

Staking is not usually necessary.

Propagation

Seed can be sown in a container under a cold frame in mid spring. Any seedlings should be ready to prick out into small pots five or six weeks later. Pot on in mid summer and plunge outdoors. Plant in their flowering positions in early or mid autumn.

Root cuttings taken in mid winter can provide good plants within a year, and this is the best method. Insert six cuttings around the edge of a 13 cm (6 in) pot and place under a cold frame. Pot up into small pots when the leaves appear in mid spring. Pot on in mid summer, plunge the pots outdoors, and then treat as for seed-raised plants.

Division of large clumps in early autumn or mid spring is a quick and easy way to obtain a few plants.

SOME POPULAR SPECIES	
E. humilis (Asia) Dark green, almost spineless leaves, with a cobwebby covering on top. Blue flowers about 2.5 cm (1 in) across, mid and late summer. 1.2–1.5 m (4–5 ft).	**E. ritro** (Southern Europe, Balkans) Divided, prickly foliage, downy beneath. Globular steely blue flowers, produced in mid summer to early autumn. 1–1.2 m (3–4 ft).

Striking heads of the globe thistle, Echinops

Epimedium

Barrenwort, Bishop's hat

H/O(M)/PS

Top *Epimedium* x *versicolor*
Bottom *E.* x *youngianum* 'Niveum'

A genus of about 20 evergreen or semi-evergreen perennials, useful as ground cover in partial shade. Although the flowers and foliage are not bold or striking, they are useful for floral arrangements.

The name is derived from *epimedion,* an ancient Greek name used by Pliny.

How to grow

Although sun is tolerated, these plants will not do well in dry, hot, sunny positions. A moist spot is ideal, and an annual mulch of moist peat or pulverised bark will help.

Remove old leaves just before the flower spikes form, usually in early spring.

Propagation

Seed is not an easy method, but might be worth considering if you need a lot of plants. Sow in containers under a cold frame in mid or late summer. Prick out the seedlings individually into small pots as they develop in the following spring (germination is sometimes erratic). Grow on under the frame for another year, then plant out in mid or late spring.

Division is much easier. This can be done in early or mid autumn or early or mid spring. Split and tease apart carefully.

SOME POPULAR SPECIES
E. grandiflorum *(E. macranthum)* (Japan) Divided, evergreen foliage. Spurred flowers about 2.5 cm (1 in) long in late spring and early summer. Colours include pink, yellow, and crimson-carmine. 23 cm (9 in). **E. perralderianum** (Algeria) Long-stalked, divided, almost evergreen, glossy leaves, colouring copper-bronze in autumn and winter. Bronzy-yellow flowers produced in mid and late spring. Good ground cover. 30 cm (1 ft). **E. x rubrum** (Garden origin) Mid green leaves, turning orange and yellow in autumn. Crimson flowers in late spring. 30 cm (1 in). **E. x versicolor** 'Sulphureum' (Garden origin) Mid green semi-evergreen leaves. Pendulous clusters of pale yellow flowers, not unlike miniature daffodils, in late spring. Vigorous. 30 cm (1 ft). **E. x youngianum** (Japan) Green leaves marked with red when young, becoming orange-red tinged in autumn. Neat growth. Showy pink flowers in mid and late spring, ivory-white flowers in the variety 'Niveum'. 25 cm (10 in).

Eranthis
Winter aconite
B/O/FS–PS

Although not really a border plant in the usual sense, these plants have been included because they bring some colour at a time when little else is in flower. Being ground-hugging, they need a place close to the front of the border, though they are even more suitable for carpeting open spaces.

The name is derived from the Greek words *er* (spring) and *anthos* (flower), reflecting the early-flowering nature of these plants.

How to grow
Well-drained but moisture-retentive soil is desirable. Plant the tubers about 5–8 cm (2–3 in) deep and 8–10 cm (3–4 in) apart, in early autumn, after soaking them overnight.

Dry tubers are sometimes difficult to start into growth, and they may take a couple of seasons to become established. Leave them undisturbed for as long as possible.

Propagation
Seed is a slow and tricky method, but can be tried if a large number of plants is required. Try to sow fresh seed, in mid or late spring. Germination is usually slow and erratic. Leave the seedlings undisturbed for two years, finally planting out the small tubers in late summer when the leaves have died down. They will not flower for another couple of years.

Divide well-established clumps, lifting in late summer. Replant the tubers without delay—root disturbance is resented, and they may take 18 months to flower again.

Top *Eranthis* x *tubergenii*
Bottom *E. cilicica*

SOME POPULAR SPECIES	
E. cilicica (Asia Minor) Deep yellow buttercup-like flowers, with a green ruff. Early or mid spring. 5–8 cm (2–3 in). **E. hyemalis** (Europe) Bright yellow, buttercup-type flowers about 2.5 cm (1 in) across, with a collar of fringed leaves. Late	winter or early spring. 10 cm (4 in). **E. x tubergenii** (Garden origin) Similar to previous two species, but with larger, sweetly scented, flowers in early and mid spring. 10 cm (4 in).

Eremurus

Foxtail lily
H/O/FS

A genus of about 50 species, taking its name from the Greek words *eremos* (solitary) and *oura* (a tail)—referring to the single flower spike.

The eremurus provide some of the giants of the border. A large group will look spectacular, but they need very careful placing among other plants if they are not to dominate and create an unbalanced effect.

Despite their large size, they make magnificent flowers for the home.

How to grow
These plants need a fertile loam. And because they produce large fleshy roots that radiate from the central crown like the spokes of a wheel not far beneath the surface, they should not be planted where you are likely to cultivate the soil around the plants. Choose a site sheltered from cold winds.

Severe winters may kill the plants. As a precaution, cover the crowns with bracken or peat in late autumn, and remove the protection in spring.

Leave the plants undisturbed for as long as possible.

Young plants may take three or more years to flower but they are more likely to establish easily.

Mulch each spring, and make sure the soil is not allowed to dry out in summer.

Cut down the stems after flowering unless seeds are required.

Propagation
Seed is a slow method and can be tricky, but the resulting plants usually grow away with much less check than divided plants.

Sow in containers under a cold frame in autumn, and bring them into warmth (16°C/60°F) in mid or late winter to germinate. Germination is often slow and erratic and may take up to eight months. Pot up and grow on for a year under the frame, then pot on in early autumn to grow on for another year. They should then be ready to plant out in mid autumn.

Division is more convenient and quicker.

Choose crowns at least two or three years old and cut down any remaining flower stems in early or mid autumn before lifting. Make sure each piece has at least one or two good buds. The plants will probably resent the root disturbance, and re-establishment is often slow.

SOME POPULAR SPECIES	
E. elwesii (Origin uncertain) Strap-shaped leaves. Spikes of fragrant pale pink flowers in late spring. There is also a white variety. 2.1–3 m (7–10 ft). **E. robustus** (Turkestan) Narrow, strap-shaped leaves up to 1.2 m (4 ft) long. Peach coloured flower spikes, late spring	and early summer. 2.4–3 m (8–10 ft). **E. stenophyllus** *(E. bungei)* (Iran, Afghanistan, Turkestan) Strap-shaped leaves. Spikes of golden yellow flowers with orange anthers, produced in early summer. 1–1.5 m (3–5 ft).

Eremurus Shelford Hybrid

Erigeron
Fleabane
H/O/FS

A genus with over 200 members, having daisy-like flowers, some rather weedy, others attractive border plants resembling dwarf Michaelmas daisies. They also make good cut flowers for arrangement in the home.

The erigerons most widely grown are the *E. speciosus* hybrids. These are usually listed simply under their varietal name, such as *E.* 'Darkest of All', *E.* 'Foerster's Liebling', *E.* 'Prosperity' and *E.* 'Rotes Meer'.

How to grow
Not fastidious about soil, but avoid any position where they are likely to suffer from lack of water in summer.

Propagation
Seed is an easy method of propagation. Sow in containers under a cold frame in mid or late spring, or germinate in warmth (16°C/60°F) in early spring. Either way prick out seedlings into pots when large enough to handle (usually after about six weeks), harden off and plunge out-doors. Plant in their flowering positions in autumn.

Division is quick and easy. If a lot of plants are required, single pieces of rooted stem can be used. Autumn or spring are suitable times for doing this.

SOME POPULAR SPECIES	
E. aurantiacus (Turkestan) Deep orange flowers produced in succession all summer. 30 cm (1 ft). **E. macranthus** (Western USA) Erect habit. Lavender-blue flowers up to 5 cm (2 in) across, mid and late summer. 60 cm (2 ft).	**E. speciosus** (Western USA) Violet ray petals surrounding a golden disc, mid and late summer. But it is mainly the hybrids that are grown—and these are in shades of pink, violet, blue, some double or semi-double. 45–60 cm (1½–2 ft).

Top *Erigeron* 'Mrs F.H. Beale'
Bottom *E.* 'Foerster's Liebling'

Eryngium
Sea holly
H/O(D)/FS

A genus of over 200 species containing mainly unusual, spiny-looking plants, many with a metallic sheen. Unless otherwise stated, the foliage is divided and spiny, although stem leaves may vary from basal leaves. The teasel-like flower heads will retain much of their colour and sheen for winter decoration if cut in their prime and dried slowly.

How to grow
A sunny position and deep, well-drained soil is desirable. The plants send down long, thick, thong-like roots.

If not already cut for decoration, take all the stems back almost to ground level in autumn.

Propagation
Seed is an easy method. Surface sow in containers in early or mid spring. Most should germinate within a month if kept at 18–21°C (65–70°F). Prick out into small pots about late spring, and grow on a few degrees cooler for three or four weeks then move to a cold frame. Harden off and plunge outdoors until early autumn, when they can be planted in their flowering positions.

Root cuttings about 6 cm (2½ in) long taken in late winter should root without much difficulty in a cold frame. Use a sandy cuttings' compost. Pot

Eryngium giganteum

up singly when the shoots show, pot on in early or mid summer, and plant in autumn after plunging outdoors for the summer.

Division is easy with tufted or clump-forming kinds such as *E. alpinum*.

SOME POPULAR SPECIES	
E. alpinum (Europe) Rounded green leaves. Metallic-blue flowers, mid summer to early autumn. 60 cm (2 ft). **E. bourgatii** (Pyrenees) Silvery-blue leaves and flowers, surrounded by long silvery bracts, mid and late summer. 45 cm (1½ ft). **E. x oliverianum** (Garden origin) Vivid blue-green deeply cut leaves on stout stems. Large, blue 'thimble'	flowers, mid summer to early autumn, 1–1.2 m (3–4 ft). **E. planum** (Eastern Europe, Asia) Dark green, heart-shaped leaves. Deep blue, globular flowers mid and late summer. 60–90 cm (2–3 ft). **E. tripartitum** (Mediterranean area) Smooth, dark green leaves. Grey-blue flowers, mid and late summer. 90 cm (3 ft).

Eupatorium
Hemp agrimony
H/O/FS–PS

A genus of over 1,000 species, being given this classical name because Mithridates Eupator, King of Pontus, discovered a poison antidote in one of the species. These are not plants for a small border, but are useful space-fillers in a large border.

Eupatorium purpureum

How to grow

Undemanding with regard to soil, though they prefer reasonably moist ground.

Despite the tall height of some species, they do not require staking. Will make large, very substantial clumps in time.

Propagation

Seed can be sown, two or three to a pot, in April under a cold frame. Thin to one in each pot, pot on in mid summer, then plunge outdoors until planting time in autumn.

Cuttings may need a bit of encouragement. Use basal, non-flowering shoots 8–10 cm (3–4 in) long in early or mid spring. Use a rooting hormone and provide warmth (13–18°C/55–65°F). Pot up when rooted, and gradually harden off under a cold frame, then treat as seed-raised plants.

Division in autumn or spring is the easiest method.

SOME POPULAR SPECIES	
E. maculatum (North America) Compact, deep lilac flowers, mid summer to late autumn. 1 m (3 ft). **E. purpureum** (North America) Purple-tinged leaves. Purple-rose flower heads	on stiff stems. 1.8–2.1 m (6–7 ft). **E. rugosum** (*E. fraseri*) (North America) Nettle-shaped foliage. Stout stems. Loose heads of fluffy white flowers, late summer to early autumn. 1–1.2 m (3–4 ft).

Euphorbia

Spurge

H/O/FS–(SD, some)

A large genus with about 2,000 species, named after Euphorbus, physician to King Juba of Mauritania. Some are strange-looking plants, but there are several very beautiful border species. Those below are typical of the border type. The true flowers are insignificant, and the attractive parts of the flowers are the petal-like bracts.

How to grow

The species here will grow in any reasonable soil. Most do best in full sun, but *E. robbiae* thrives in shade.

Propagation

Seed can be sown indoors in early spring—a temperature of 18–21°C (65–70°F) is about right for most. Pot up and plunge outdoors for the summer. Plant out in autumn.

Cuttings of 8 cm (3 in) long basal shoots will root without much trouble. Pot up when rooted, plunge outdoors for the summer, and plant in autumn.

Division, in early or mid autumn, or in spring, is the easiest method.

SOME POPULAR SPECIES	
E. cyparissias (Europe) Pale, linear leaves. Yellow flower heads from late spring to mid summer. Creeping habit. Makes an effective ground cover, but invasive. 25 cm (10 in). **E epithymoides** (*E. polychroma*) (Eastern Europe) Sulphur-yellow flower heads on mounds of fresh green leaves in mid and late spring. Neat habit. 45 cm (1½ ft). **E. griffithii 'Fireglow'** (Himalayas) Lanceolate leaves with	pale pink midribs. Orange-red flower heads in late spring and early summer. 60 cm (2 ft). **E. myrsinites** (Southern Europe) Trailing stems with glaucous grey, evergreen leaves. Wide heads of lime-green flowers in spring. 15 cm (6 in). **E. robbiae** (Asia Minor) Rosettes of dark green leathery leaves, topped with showy heads of yellowish green flowers in early and mid summer. 45–60 cm (1½–2 ft).

Euphorbia epithymoides

Filipendula

H/O(M, most)/FS–PS

A genus of ten hardy herbaceous perennials, taking its name from the Latin words *filum* (a thread) and *pendulus* (hanging), referring to the thread-like roots by which the tubers of *F. vulgaris* are attached.

How to grow

Ordinary, well-drained soil is suitable, but most do best in moist ground. Of those listed here, only *F. vulgaris* prefers fairly dry conditions (it also does best in an alkaline soil). All of them respond well to feeding.

Mulch the plants each spring. Cut the stems down in mid autumn.

Propagation

Seed-raised plants are successful and moderately easy. Sow in warmth (10–13°C/50–55°F) in late winter or early spring. Prick out into small pots and move out to a cold frame. Pot on in mid summer, harden off and plunge outdoors. Set the young plants out in their flowering positions in early autumn (but do not expect flowers the following year).

Division is quick and easy. Cut down old flowered stems and lift large clumps to divide in autumn or spring.

Filipendula palmata

SOME POPULAR SPECIES	
F. palmata (Siberia) Divided foliage. Feathery heads of small pink flowers in mid summer. 75 cm (2½ ft). **F. rubra** *(F. venusta)* (Eastern USA) (Queen of the prairie) Pinnate, dark green foliage. Wide heads of small pink flowers on stiff, branching stems in mid and late summer. 1.2–1.5 m (4–5 ft). **F. ulmaria** (Meadowsweet, queen of the meadows) Pinnate	foliage dark green above, whitish beneath, topped with flattened heads of creamy-white fragrant flowers in summer. 'Aurea' has golden foliage (do not allow this one to flower). 30–45 cm (1–1½ ft). **F. vulgaris** *(F. hexapetala)* (Europe) (Dropwort) Dark green ferny foliage. Pink-tinged white flowers in early and mid summer. It is usually the double form that is grown. 60 cm (2 ft).

Gaillardia

Blanket flower

H/O(D)/FS

A genus of 28 species, commemorating M. Gaillard de Marentonneau, a French patron of botany. Only one species is generally grown as a border plant though there is also an annual that is sometimes grown. The flowers are large and long-stalked, and suitable for cutting.

Gaillardia 'Wirral Flame'

How to grow

Will grow in almost any soil in full sun. Will do well on dry soil, but this is by no means essential.

The support of twiggy sticks will be necessary.

Dead-head to prolong flowering, and to keep the plants looking tidy.

Propagation

Seed is a popular method if you do not want a named variety. Sow indoors in warmth (16°C/60°F) in late winter. Prick out into small pots and grow on a few degrees cooler. Harden off and plant out in late spring where they are to flower. They should flower the same year.

Root cuttings root without difficulty if placed in a cold frame in late winter. Make them about 8 cm (3 in) long, and pot up about mid spring. Pot on in summer, harden off and plunge outdoors until planting time in autumn.

Division in autumn or spring is easy.

SOME POPULAR SPECIES	
G. aristata *(G. grandiflora)* (North America) Grey-green lanceolate leaves. Lax habit. Rayed, daisy-type flowers	5–10 cm (2–4 in) across, red or orange usually tipped yellow, all summer. 45–75 cm (1½–2½ ft); 'Goblin' is 30 cm (1 ft).

Galtonia

Summer hyacinth

B/O/FS–PS

A small genus of only four species, commemorating Sir Francis Galton, a 19th century anthropologist. Only one species is in general cultivation, and this is a bold plant, especially in an established clump, for the middle or back of the border, or among shrubs.

How to grow

Plant the bulbs in groups of about five, 15 cm (6 in) deep, in mid autumn or early or mid spring. Leave them undisturbed until crowded, then they can be lifted and the bulbs and offsets replanted.

Propagation

Seed is a slow, challenging method. Offsets are more dependable.

Offsets are sparingly produced, but can be removed if the bulbs are lifted in early autumn. Replant immediately, or pot up the bulblets and overwinter under a cold frame, then plunge outdoors in late spring.

SOME POPULAR SPECIES	
G. candicans (South Africa) Strap-shaped leaves about 60 cm (2 ft) long. Stiff	spikes of large nodding white bells marked green produced in summer. 1.2 m (4 ft).

Galtonia candicans

Gentiana
Gentian
H/O/FS–PS

A large genus of about 400 species of annuals and herbaceous perennials—mostly rock garden plants, but a few species are good border plants.

The genus is named after Gentius, King of Illyria, who first used one of the plants medicinally.

Gentiana lutea

How to grow
It is worth adding plenty of grit to the planting area if the ground is heavy, while leafmould or damp peat will also help if the soil is likely to be dry. They like a well-drained root-run in winter but plenty of moisture in summer.

Although some species of gentian require an acid soil, those listed below are less demanding.

Most enjoy sun, but *G. asclepidea* is best in partial shade.

Leave the clumps undisturbed until they begin to deteriorate. They dislike being moved, much less divided.

Propagation
Seed needs to be fresh, and even then propagation can be tricky. Sow in early autumn in containers in a cold frame, and be prepared for germination to be erratic and take up to 18 months. Prick out the seedlings into 10 cm (4 in) pots, harden off and plunge outdoors for the summer. Plant in autumn.

Cuttings can also be tricky and a bit of a challenge. Use 4–5 cm (1½–2 in) long basal shoots, and root them in a sandy compost under a cold frame in mid spring. Pot up singly, harden off and plunge outdoors for the summer. Give winter protection, and finally plant out in early or mid spring.

Division in early spring is the easiest method.

SOME POPULAR SPECIES	
G. asclepiadea (Europe) (Willow gentian) Willow-like leaves in opposite pairs. Intense blue tubular flowers clustered on graceful arching stems in mid and late summer. 60 cm (2 ft).	**G. lutea** (Europe) (Yellow gentian) Stiff, erect habit, with tall, stout stems and large leaves. Whorls of yellow flowers in mid and late summer. 1.2 m (4 ft).

Geranium
Crane's-bill
H/O/FS–PS

A large genus of about 400 species of hardy herbaceous perennials. Not to be confused with the bedding 'geraniums', which are really pelargoniums. Some of the geraniums used in gardens are vigorous and rather rampant and aggressive. This can be useful if you want ground cover, but can be a problem in a mixed border, so choose carefully.

Geranium psilostemon

How to grow

Good drainage is the chief cultural requisite, otherwise they will tolerate almost any type of soil. If the ground is too rich, lush foliage may be produced at the expense of flowers.

Cut back old flowering stems to just above ground level to keep the plants compact and perhaps encourage a further flush of flowers.

Propagation

Seed can be sown in containers in a cold frame, in either autumn or early spring. Alternatively sow in warmth (10–13°C/50–55°F), in late winter. Prick out singly in to small pots, harden off in summer and plunge outside. Plant in early autumn or spring. Division in autumn or spring is easy, but do not replant too deeply.

SOME POPULAR SPECIES			
G. endressii (Pyrenees) Palmate, deeply lobed leaves. Evergreen. Pale pink flowers, lightly veined red, all summer. 'Wargrave Pink' is a good form with clear pink flowers. Good ground cover. 30–45 cm (1–1½ ft).	**G. ibericum** *(G. platypetalum)* (Caucasus, Iran) These are the names of distinct plants, but horticulturally they have become confused. Violet-blue flowers, 2.5 cm (1 in) across, in mid and late summer. 60 cm (2 ft).	**G. pratense** (Northern Europe) (Meadow crane's bill) Lobed, deeply divided leaves. Blue or violet-blue flowers from mid summer to early autumn. There is a white variety, and a double blue form. The hybrid 'Johnson's Blue' is	outstanding. 45–60 cm (1½–2 ft). **G. psilostemon** *(G. armenum)* (Armenia) Clumps of broad palmate leaves that colour in autumn. Magenta flowers with dark centres, in early and mid summer. 75 cm (2½ ft).

Geum

Avens

H/O/FS–PS

A genus of 40 species of hardy herbaceous perennials. Typical 'cottage garden' plants, and useful for cutting.

The most popular geums are varieties of *G. chiloense*, but are most likely to be sold by varietal name alone (e.g. *Geum* 'Mrs Bradshaw').

How to grow

Ordinary but fertile soil. Partial shade is tolerated but full sun will produce better plants. Be prepared to water in very dry weather.

G. chiloense is likely to need the support of twiggy sticks in an exposed position.

The plants are best lifted and divided every two or three years.

Propagation

Seed provides a reliable method of propagation. Sow in containers under a cold frame from late spring to mid summer. Prick out into small pots, harden off, and plunge outdoors. Give winter protection, and plant out finally in mid or late spring.

Alternatively sow in warmth (18–21°C/65–70°F) in late winter or early spring. Prick out into small pots, harden off, and plunge outdoors for the summer. Plant out in early autumn.

Division is easier. This is best done in early or mid spring.

SOME POPULAR SPECIES	
G. x borisii (Garden origin) Leafy plants topped with orange-scarlet single flowers on long, wiry stems, from late spring until late summer. Can be used as a ground cover. 30 cm (1 ft).	**G. chiloense** (Chile) The species itself is not grown, but its varieties are: 'Mrs Bradshaw' is a semi-double scarlet, 'Lady Stratheden' is a double yellow. 45–60 cm (1½–2 ft).

Geum 'Lady Stratheden'

Gypsophila
Chalk plant
H/AK(D)/FS

A genus of 125 species of hardy annuals and herbaceous and evergreen perennials. The name is derived from the words *gypsum* (gypsum) and *philos* (friendship), alluding to the fact that some species like growing on gypsum, alkaline, rocks. The individual, usually white, flowers are small and not particularly pretty, but the sheer number of them make the whole plant a frothy mass of bloom. A traditional florists' plant, the flowers lasting well in water and mixing with other plants in an arrangement. In the border they act as a good foil for other plants.

Gypsophila paniculata

How to grow
Gypsophilas do best on alkaline soils. The other requirement is a deep root-run (sometimes difficult on chalky soils as these are often shallow).

The plants produce deep, thong-like roots and resent being moved.

Propagation
Seed is a fairly easy method of propagation. Double varieties are likely to produce singles too. Sow in containers under a cold frame in early or mid spring. They should be ready for pricking out after about a month. Pot on in mid summer, harden off and plunge outdoors. Give winter protection, and plant out in mid spring.

Root cuttings 6 cm (2½ in) long, taken in late winter and rooted in a cold frame, usually root readily.

Cuttings are not easy. Use 8 cm (3 in) long basal shoots, in mid or late spring—they may root in about a month. Use a sandy compost, and pot up singly once rooted; pot on in mid or late autumn. Plunge outdoors for the summer, give winter protection, then plant in spring.

SOME POPULAR SPECIES	
G. paniculata (South Eastern Europe, Siberia) Grey-green lanceolate to linear leaves. Small white flowers in large, loose	panicles. All summer. 'Bristol Fairy' has double flowers. There are also pink varieties. 1–1.2 m (3–4 ft).

Helenium
Sneezeweed
H/O/FS

A genus of 40 herbaceous perennials and annuals, named after Helen of Troy—according to legend the flowers sprang from her tears. The daisy-type flowers are good for cutting, lasting well and having long, straight stems.

How to grow
Heleniums grow well in almost all soils and plants will probably need staking.

Cut back the flowered stems once they have finished. It will make the plants look much tidier and may encourage further flowers.

H. autumnale 'Wyndley' and 'Moerheim Beauty'

Propagation

Seed sown in containers under a cold frame in mid or late spring should be ready for pricking out into small pots after about a month. Pot on in mid summer and plunge outdoors. Plant out in autumn.

Division is easy and the normal method of propagation. It is best done in autumn when the plants have finished flowering, or alternatively in early spring.

SOME POPULAR SPECIES	
H. autumnale (Canada, USA) Lanceolate leaves. Normally yellow flowers about 3–4 cm (1–1½ in) across. It is mostly its varieties that are grown, however, and these have	flowers in shades of red and orange as well as yellow. Flowering period depends on variety, within the range mid summer to mid autumn. 60–120 cm (2–4 ft), according to variety.

Helianthus

Perennial sunflower

H/O/FS

The annual sunflower, known to almost everyone, also has its perennial relatives. The perennial sunflowers are much more suitable than the annuals for general flower borders, and although less massive in height they are still very bold and attractive, if sometimes rather spreading. They are suitable for cutting.

The genus, containing 55 species, takes its name from the Greek *helios* (sun) and *anthos* (a flower).

How to grow

Cultivation is easy. They grow well in almost any soil. Best in full sun.

Remove dead flowers. Cut flowering stems almost to ground level after flowering.

Propagation

Seed is an easy method if you can obtain a suitable supply. Sow in containers under a cold frame in early or mid spring. Better still, sow in warmth (13–16 °C/55–60 °F). They should be ready to prick out singly into 10 cm (4 in) pots in three to five weeks. Grow on under a cold frame, pot on into 13–15 cm (5–6 in) pots in summer, and plunge outdoors. Plant out in early or mid autumn.

Division in early autumn or in mid spring is easy. Cut down any remaining stems, fork up the clumps and pull away rooted offsets. These can then be replanted.

SOME POPULAR SPECIES	
H. atrorubens *(H. sparsifolius)* (USA) Rough, hairy ovate leaves. Yellow flowers with red central disc, early autumn. 1.2–2.1 m (4–7 ft). **H. decapetalus** *(H. multiflorus)* (USA) Rough, sharply toothed ovate leaves. Pale yellow flowers about 5 cm (2 in) across, mid summer to early autumn. There are	several varieties, including semi-double and double forms, such as 'Loddon Gold' (double). 1.2–1.5 m (4–5 ft). **H. salicifolius** *(H. orgyalis)* (USA) Long, almost grass-like foliage. Sprays of small bright yellow flowers produced in late summer and early autumn. 1.8–2.4 m (6–8 ft).

Helianthus 'Loddon Gold'

Heliopsis
H/O/FS

A small genus of a dozen species, including both annuals and perennials. The name is derived from the Greek *helios* (sun) and *opsis* (like), referring to the flowers. They are stiff, rather inelegant plants with the flowers on erect, branching stems. They are good for cutting.

How to grow
Easy to grow provided the ground is reasonably fertile. Water freely in dry weather.

Cut down the stems almost to ground level in late autumn and divide the plants every third year.

Propagation
Seed is little used, but can be sown in warmth (16 °C/60 °F) in early or mid spring. After potting up, hardening off, and plunging outdoors for the summer, they should be ready for planting in early autumn.

Division is the usual method, and can be done in autumn or early or mid spring.

Heliopsis 'Light of Loddon'

SOME POPULAR SPECIES	
H. scabra (North America) Rough, lanceolate leaves and single daisy-type yellow flowers about 8 cm (3 in) across and produced	in mid and late summer. There are several varieties, including semi-doubles and doubles such as 'Golden Plume'. 1–1.5 m (3–5 ft).

Helleborus
Hellebore
H/O(M)/PS–SD

The best-known species are *H. niger* (Christmas rose) and *H. orientalis* (Lenten rose), mainly because they bloom when there is little else in flower. They are worth planting for that reason alone, but some of the other species, such as *H. foetidus*, are perhaps more useful general border plants.

The name is derived from two Greek words: *helein* (to kill) and *bora* (food)—some species being poisonous.

How to grow
Hellebores will thrive in places that never receive direct sunlight, but they are also happy in partial shade.

A soil with plenty of humus, not likely to dry out, is best. It should never be waterlogged.

Propagation
Seed is slow, difficult, and erratic in germination.

Helleborus orientalis

Sow in containers of sandy compost in mid summer and place under a cold frame. Prick out any seedlings that show in the first year—singly into small pots. Overwinter under the frame and plant out in mid spring.

Leave the container of seeds in the frame for a second winter, then bring it into warmth (16–18°C/60–65°F) in late winter—more seeds should germinate within the next couple of months. Prick them out, grow on cooler, and plunge in a cold frame. Uncover for the summer, but protect in winter, then plant out in spring.

Division is easy but is likely to give the plants a severe check. Cut up into pieces, each with three or four leaves. Divide *H. niger* and *H. orientalis* in spring, the remainder in autumn.

SOME POPULAR SPECIES	
H. atrorubens (South Eastern Europe) Dark green deciduous leaves. Plum-purple cup-shaped flowers from mid winter to early spring. 25 cm (10 in). **H. foetidus** (Europe) (Stinking hellebore) Upright, dark evergreen leaves. Light green flowers, mid to late spring. 60 cm (2 ft). **H. niger** (Central and Southern Europe) (Christmas rose) Dark,	leathery, evergreen lobed leaves. White flowers about 5 cm (2 in) across, from early winter to early spring. 30–45 cm (1–1½ ft). **H. orientalis** (Greece, Asia Minor) (Lenten rose) Evergreen or semi-evergreen lobed leaves. Two or three flowers on each stem, varying from white to plum purple, some spotted maroon inside, late winter to mid spring. 30 cm (1 ft).

Hemerocallis

Day lily

H/O(M)/FS–PS

A genus of 20 hardy herbaceous perennials, taking their name from the Greek words *hemero* (a day) and *kallos* (beauty), a reference to the short but beautiful life of the flowers. Individual flowers open for only a day, but the succession of blooms means that the plants appear to be in flower for a long period. It is mainly the numerous hybrids that are grown.

The day lily, Hemerocallis

How to grow

Best in moist soil, and they make attractive waterside plants, but do not let this deter you from using them in an ordinary border.

Leave the clumps undisturbed for as long as possible.

Propagation

Seed is an excellent way to start a collection of these plants. By sowing in warmth (18°C/65°F) in mid winter, they should flower the first year. Prick out into small pots and grow on a few degrees cooler. Pot on in late spring, harden off, and plant out in the border in early summer.

Division is an easy, no-fuss way to obtain a few extra plants. Cut down the old flowered stems, lift and split up the fleshy roots. It can be done in autumn, but in a wet season it is best to delay until spring. Cut the leaves back by half after replanting.

SOME POPULAR SPECIES	
H. citrina (China, Japan) Strap-shaped leaves. Lily-like lemon-yellow flowers, 10 cm (4 in) across, in mid and late summer. 1–1.2 m (3–4 ft). **H. fulva** (Japan, Siberia) Strap-shaped leaves, red flowers tinged apricot produced in summer. 90 cm (3 ft).	**Garden hybrids** (Garden origin) Bold, strap-shaped leaves. Trumpet-shaped, lily-like flowers in shades of yellow, orange, red and pink, some attractively marked. There are many varieties, all flowering in summer. 45–90 cm (1½–3 ft).

Hesperis
Sweet rocket
H/O(M)/FS

Hesperis matronalis

A genus of 30 herbaceous perennials and biennials. The genus takes its name from the Greek *hesperos* (evening), probably because the flowers of some species are fragrant in the evening.

How to grow
Best in a moist, but not heavy or waterlogged soil. Otherwise easy to grow.

Cut out flowering spikes once they have finished blooming.

Propagation
Seed is the preferred method of raising this perennial, which is best treated as a biennial.

Sow in mid spring in a sunny position in open ground (or under cloches or in a frame if possible). Prick off the seedlings into a nursery bed in late spring or early summer. Plant in flowering positions in autumn.

Division of established plants in autumn or spring is possible but not always satisfactory.

SOME POPULAR SPECIES	
H. matronalis (Southern Europe, Western Asia) Dark green lanceolate leaves. The spikes of	white, mauve or purple flowers are produced in early summer. 60–90 cm (2–3 ft).

Heuchera
Coral bells, coral flower
H/O/FS–PS

A genus of 50 hardy herbaceous perennials, named in honour of Professor J. H. Heucher (1677–1747), a German professor of medicine and a botanist. The flowers are very pretty cut for indoor decoration.

How to grow
Not difficult to grow provided the soil is not too heavy. Waterlogging is likely to prove fatal for the plants. Partial shade is better than full sun.

Plant with the crowns nestling on the ground and their long brown roots well spread out. The

Heuchera 'Scintillation'

crowns sometimes tend to rise out of the ground after a few years. If this happens, mulch liberally or lift, divide, and replant.

Propagation
Seed sown in mid spring in containers under a cold frame should be ready for pricking out into small pots in about a month. Pot on in summer and plunge outdoors until autumn, when they can be planted.

Division is a popular method. Lift and pull away rooted offsets in autumn or mid spring. Replant the divisions 3–4 cm (1–1½ in) deeper than they were before.

SOME POPULAR SPECIES	
H. sanguinea (Arizona, Mexico) Clumps of dark green, rounded, sometimes marbled, leaves. Panicles	of dainty pink, red, or white bell-shaped flowers on thin stems from mid summer to early autumn. 30–60 cm (1–2 ft).

Hosta
Plantain lily
H/O(M)/FS–PS

A genus of about 20 species of hardy perennial, commemorating Nikolaus Thomas and Joseph Host, Austrian botanists of the late eighteenth and early nineteenth centuries. They are ground cover plants once established, and are impressive in bold groups in the herbaceous border. Hostas also associate well with water.

Although some do have quite attractive flowers, they are really best regarded as foliage plants.

How to grow
Hostas do well in most soils, provided they are kept moist. The plants do well in full sun, will even tolerate shade, but are perhaps at their best in partial shade. Variegated varieties have the best colouring in light shade, but bear in mind that the variegation is likely to fade anyway with the advancing season.

Slugs and snails love hostas. Unless you take precautions against them, much of the beauty of these plants may be lost.

Propagation
Seed can be successful, but bear in mind that variegated varieties will show some variance and will probably include some plain-leaved progeny. Sow in warmth (10–13 °C/50–55 °F) in early spring. It should germinate in one to three months. Prick out into small pots and grow on under a cold frame, potting on in summer before plunging outdoors. Give winter protection and plant out in spring.

Division of established crowns in spring or autumn is the quickest and easiest method. Make sure each segment of the crown has at least one 'eye'.

SOME POPULAR SPECIES	
H. albo-marginata *(H. sieboldii)* (Japan) White-edged leaves. Lilac flowers with violet stripes, mid and late summer. 45 cm (1½ ft). **H. crispula** (Japan) Dark green leaves with white margins. Mauve flowers, mid and late summer. 60 cm (2 ft). **H. fortunei** (Japan) Glaucous grey-green leaves. Lilac-blue flowers, in mid summer. There are several very attractive variegated forms, such as	'Aureo-marginata' and 'Albopicta'. 60–75 cm (2–2½ ft). **H. lancifolia** (China, Japan). Narrow, lanceolate green leaves. Pale lilac flowers, very freely produced in mid and late summer. 45–60 cm (1½–2 ft). **H. sieboldiana** (Japan) Glossy, prominently veined leaves, blue-green in the variety 'Elegans'. Lilac-white flowers in late summer. 60 cm (2 ft).

Hosta, a useful plant for light shade

Incarvillea

H/O/FS

A genus of about 14 species of herbaceous perennials, commemorating Pierre d'Incarville (1706–57), a French Jesuit missionary to China.

The plants have a distinctive appearance that does not blend easily with most other herbaceous border plants, but they undoubtedly add interest.

Incarvillea delavayi

How to grow
These plants need care if they are to succeed. It is best to establish young plants (seedlings about a year old), rather than attempt to start from dry roots (which you may be able to obtain from bulb merchants or seedsmen). Once established, leave undisturbed for as long as possible. The big, fleshy, tuberous roots resent disturbance.

A deep, rich, well-drained soil and a sunny position will produce the best results.

Propagation
Seed is not an easy method, but you should be successful. Sow in warmth (18–21°C/65–70°F), a few seeds in each pot of soil-based, sandy compost in early or mid spring. Germination should take about a month. Thin to one seedling in each pot and gradually harden off under a cold frame. Leave the plants in the frame to overwinter, then plant out in spring.

Division is possible but difficult. The crowns are hard to split, and this is best attempted in mid autumn or mid spring, chopping into segments each with two or more 'eyes' and adequate roots.

SOME POPULAR SPECIES	
I. delavayi (Western China, Tibet) Dark green, deeply divided foliage. Heads of deep pink trumpet-shaped flowers, early and mid summer. 45 cm (1½ ft).	**I. grandiflora** (Western China) Entire but toothed dark green leaves. Heads of cerise-pink trumpet flowers in early and mid summer. 30 cm (1 ft).

Inula

H/O/FS

A genus of about 200 species, ranging from alpines to plants 1.5 m (5 ft) high. The border plants are showy, with yellow, daisy-like flowers.

Inula is an ancient Latin name.

How to grow
Inulas do quite well on heavy soil, but given a sunny position are happy on most soils.

Inula ensifolia

Propagation

Seed can be sown in warmth (16–18 °C/60–65 °F) in late winter. They should germinate in three or four weeks. Prick out into small pots and grow on at about 10 °C (50 °F). Move to a cold frame in mid spring, pot on about a month later, then harden off and plunge outside for the summer. Plant in position in the border in early autumn.

Division is the most popular method of increasing inulas and it can be done in either autumn or spring.

SOME POPULAR SPECIES	
I. ensifolia (Caucasus) Narrow leaves. Large golden-yellow disc-shaped flowers, early and mid summer. 23 cm (9 in). **I. hookeri** (Himalayas) Oblong-lanceolate leaves. Bushy habit. Pale yellow flowers in late summer and early autumn. 45–60 cm (1½–2 ft).	**I. magnifica** (Caucasus) Long hairy leaves. Sturdy stems with large rayed bright yellow narrow-petalled flowers. Mid and late summer. 1.8 m (6 ft). **I. orientalis** (Caucasus) Orange-yellow flowers with very narrow, wavy petals. Summer. 60 cm (2 ft).

Iris (border type, pogoniris)
Bearded iris
H/AK/FS–PS

This large genus, with over 300 species and innumerable varieties, is as colourful as the name — taken from the Greek *iris*, a rainbow — suggests. Popular bulbous species appear in the next entry; those described here have rhizomes. The plants included are only representative of these very diverse plants. Unfortunately all species tend to have a relatively short flowering season, but the period of interest can be extended by choosing species to follow on in succession.

The irises described here are popularly known as bearded iris, having a narrow, hairy 'beard' in the centre of the upper part of the falls.

The true species are seldom grown, and it is the many hybrids that are of garden value.

How to grow

Plant the rhizomes in mid autumn in a sunny position. Light, limy soil suits them well.

Do not bury the rhizome completely — leave the top half exposed. Plant in groups, spacing about 25 cm (10 in) apart. Staking should not be necessary.

Propagation

Seed offers a sensible method of building up a collection, but it is a slow business. Sow fresh seed in 13 cm (5 in) pots under a cold frame in late summer or early autumn, pre-soaking in

Border of bearded iris varieties

tepid water for 24 hours. In late winter move them into warmth 16–21 °C (60–70 °F). Germination is erratic and may take from one to eighteen months. Leave the seedlings until mid summer, then pot up individually into 13 cm (5 in) pots. Grow on for a year under the cold frame, then plant out. You are likely to wait a further year or two for flowers.

Division of the rhizomes is the usual method. This should be done in mid summer, after flowering. Lift clumps and cut off healthy pieces of rhizome 10–15 cm (4–6 in) long with one or two fans of foliage.

SOME POPULAR SPECIES	
I. pumila hybrids (Garden origin) (Dwarf bearded iris) Flowers 8–10 cm (3–4 in) across, in late spring. 10–25 cm (4—10 in). **I. germanica hybrids** (Garden origin) (Intermediate bearded iris) Flowers 8–10 cm (3–4 in) across, in late spring. 25–75 cm (10–30 in).	**I. pallida hybrids** (Garden origin) (Tall bearded iris) Flowers 10–15 cm (4–6 in) across, late spring or early summer. 75–150 cm (2½–5 ft). All the above irises have fans of sword-shaped leaves and flowers in the usual iris colours of blues, yellow, and pinks.

Iris (bulbous type)

B/O/FS

Top *Iris xiphium, Spanish iris*
Bottom *Mixed varieties of I. reticulata*

The bearded irises, perhaps the most popular group of irises for the herbaceous border, were described on the previous page. The bulbous irises most widely planted are the early-flowering types, but the summer-flowering border irises are also rewarding to grow, either as groups for decoration or as cut flowers for the home. Although these are listed here as *I. xiphium* hybrids, you will find them in shops and bulb catalogues simply as Dutch, English, or Spanish irises.

How to grow

All the bulbous irises described below are easy to grow and can be left in the ground to form large clumps. Plant in bold groups for impact—twos and threes are likely to look stiff and 'awkward'.

Plant *I. danfordiae* and *I. reticulata* about 5 cm (2 in) deep, the summer-flowering types 10–15 cm (4–6 in), in autumn.

Propagation

Division of the bulbous offsets is easier than seed, and they are produced quite freely.

Lift mature bulbs in mid or late summer, after the foliage has died down. Each bulb usually splits to reveal two or more bulbs along with a number of offsets. Plant out the largest bulbs in a bed to flower the following year. Plant the smaller offsets in a nursery bed to grow on for a year or two before planting out in their flowering positions during late autumn.

SOME POPULAR SPECIES	
I. danfordiae (Eastern Turkey) Delicately fragrant yellow flowers, opening as the leaves begin to develop, in late winter. 10 cm (4 in). **I. reticulata** (Russia, Caucasus, Northern Persia) Narrow, ribbed leaves develop while the flowers are open in late winter and early spring. There are several varieties, in shades of blue and purple. 15 cm (6 in). **I. xiphium hybrids** (France, Portugal,	Southern Spain, North West Africa) (Dutch, English, Spanish iris) Dutch irises flower first, in early summer; Spanish irises flower about a fortnight later, followed by the English irises, which bloom in mid summer. Colours include yellow, blue and purple shades, and white. All look similar, but the English type does not include yellows and the flowers are a little larger than the others. 30–60 cm (1–2 ft).

Kniphofia
Red hot poker, torch lily
H/O/FS

A genus of about 75 species, some of which are invaluable for the herbaceous border or where a striking individual plant is required. It is mainly the hybrids and varieties that are grown, but some of the species, such as *K. galpinii*, are charming border plants.

By selecting suitable species and varieties it is possible to have kniphofias in flower from early summer to mid autumn.

How to grow
Kniphofias will grow in almost any soil provided drainage is good. They will respond well to heavy watering whenever the weather is dry in summer. Wet, badly drained soils are likely to lead to winter losses.

Make sure the roots are well spread out when planting.

On heavy soil or in cold districts cover the crowns with straw or bracken to protect the fleshy crowns. Mulch each spring.

Remove the flower spikes as they fade, and in late autumn bundle the leaves together over the crown to protect it against excessive winter moisture.

Propagation
Seed sown under a cold frame in early or mid spring will flower the following year. Sow in containers and prick out singly into small pots when they are ready, about a month later. Pot on in mid summer, harden off, and plunge outdoors for the summer. Give winter protection, then plant out in spring.

Alternatively sow in warmth (16–18°C/60–65°F) in mid winter. Prick out into small pots then grow on a few degrees cooler. Pot on, harden off and plant out in late spring. Some varieties may flower the same year.

Division is easy and the main method. Do this in mid spring, and be careful not to damage the fleshy underground stems if possible when lifting the crowns.

Kniphofia 'Springtime'

SOME POPULAR SPECIES	
K. caulescens (South Africa) Forms a large clump of broader, shorter, stiffer, foliage than most species. The leaves are also glaucous like a carnation's. Flowers pinky or salmon-tinted at first, changing to yellow, in autumn. 1.2–1.5 m (4–5 ft). **K. galpinii** (Transvaal) Thin, grassy foliage drooping at the tips. Flame orange pokers early and mid autumn. 45 cm (1½ ft). **K. nelsonii** (Orange Free State) Narrow leaves. Bright scarlet flowers sometimes tinged orange, early and mid autumn. 45 cm (1½ ft).	**K. uvaria** (South Africa) Big, torch-like heads of drooping tubular flame-red florets produced in late summer and early autumn. 1.2–1.5 m (4–5 ft). **Garden hybrids** (Garden origin) Most kniphofias sold are hybrids, with heights generally ranging from 60–150 cm (2–5 ft), and 'pokers' in various shades of orange and yellow. Seedlings are likely to be unnamed, but these are still well worth considering. There are many named varieties, but as flowering time and heights can vary considerably check these points before buying.

Lamium

Dead nettle
H/O/PS–SD

A genus of perhaps 50 species, taking its name from the Greek *laimos* (a throat), in reference to the appearance of the flowers. The common name reflects the resemblance of these plants to the young growth of nettles, but they have no stings and seldom exceed 45 cm (1½ ft). They make good ground covers for shady sites and poor soil. The one problem is that they tend to be invasive.

Lamium maculatum

How to grow

Really easy plants, tolerating most soils and thriving in partial or full shade, although *L. orvala* will also do well in full sun.

If used as ground cover, trim over the plants with shears after flowering, to help produce dense leaf growth.

Propagation

Division is the normal method of increase, and is best done in early or mid autumn, or in early or mid spring. To stimulate fresh new growth for division, clip over the plants with shears in late summer or early autumn, taking the growth back almost to ground level.

SOME POPULAR SPECIES	
L. galeobdolon (Europe) (Yellow archangel) Yellow flowers in whorled spikes, late spring and early summer. 'Variegata' has white variegation. Good ground cover. Rampant. 25 cm (10 in). **L. garganicum** (Southern Europe) Compact, non-rampant plant. Deep green leaves. Pink to red flowers in early and mid summer. Needs a moist position. 15 cm (6 in).	**L. maculatum** (Europe) Close tufts of green and white variegated foliage. Purple flowers, late spring and early summer. 'Chequers' has marbled leaves. 'Beacon Silver' has silvery white foliage. 'Album' has white flowers. 10–15 cm (4–6 in). **L. orvala** (Italy, France) (Giant dead nettle) Pink to purple flowers produced in late spring and early summer. Not invasive. 60 cm (2 ft).

Lathyrus

Everlasting sweet pea
H(CL)/O/FS

A genus of 130 species, mostly climbers. Although lacking the size of flower and colour range of the annual sweet pea, the perennial species are useful climbers for the back of the border or against a fence.

Like the sweet pea, the perennial species make good cut flowers.

How to grow

Perennial sweet peas will thrive in any fertile, well-drained soil, in full sun.

Suitable support may be needed, and tall, bushy twigs or branches for them to scramble through at the back of the border are ideal.

Dead-head the plants unless seed is required.

Lathyrus vernus

Propagation

Seed is easiest, and for some species the only practical method of increase. Sow in a greenhouse (13–18 °C/55–65 °F) or cold frame in early or mid spring, first chilling the seeds for two or three weeks in a fridge at about 0 °C (32 °F), then soaking in tepid water for 12 hours. Sow one or two seeds in each 8 cm (3 in) pot. When they germinate thin to leave one plant to each pot, then harden off and plunge outdoors for the summer. Plant in autumn, where they are to flower (in cold districts it is best to overwinter the pots in a frame and plant out in spring).

Division is difficult, but might work for *L.*

latifolius. Do it in March and be careful not to let the roots dry out.

SOME POPULAR SPECIES	
L. grandiflorus (Southern Europe) Rosy-red flowers, some shading to purple. Blooms often in pairs but not sprays or spikes. Freely produced throughout the summer. 1.2–1.8 m (4–6 ft). **L. latifolius** (Europe)	Vigorous. Usually rose-purple flowers, but also shades of red and violet as well as white. Mid to late summer. 1.8–2.4 m (6–8 ft). **L. vernus** (Europe) Non climbing, purple or blue flowers, late spring. 30 cm (12 in).

Liatris

Blazing star, gayfeather

H/O/FS

A genus of about 40 species of hardy herbaceous plants. The derivation of the name is unknown.

Most spiky flowers open their lowest blooms first and proceed upwards, lengthening the upper part of the spike more or less in unison. The liatris are unusual in that the spikes grow to their full length while still in the bud stage, then open from the tip first, working downwards.

They make good cut flowers.

How to grow

Well-drained soil is essential. Waterlogged soil in winter will lead to losses. They need abundant water during the summer, however, so be prepared to water freely in dry weather. Incorporate plenty of well-rotted compost or manure when planting. Mulch annually.

Propagation

Seed can be sown in early spring, either under a cold frame or in a greenhouse (13–16 °C/55–60 °F). Seedlings will be ready to prick out into small pots after about a month. Grow on under the frame, pot on, then harden off and plunge outdoors. Plant in early autumn.

Division is a popular and easy way to propagate these plants. As new growth breaks in early or mid spring, lift an established clump with a fork, pull off and plant rooted offsets.

SOME POPULAR SPECIES	
L. callilepis (USA) Linear leaves. Stiff leafy spikes of carmine fluffy flowers produced from mid summer to early autumn. 90 cm (3 ft). 'Kobold' (deep lilac flowers) is dwarfer— 60 cm (2 ft).	**L. graminifolia** (Eastern USA) Sparse, narrow leaves. Purple flowers on slender spikes. 60–90 cm (2–3 ft). **L. spicata** (USA) Rosy-purple truncheons of bloom, in early autumn. 60–90 cm (2–3 ft).

Liatris 'Kobold'

Ligularia
Leopard plant
H/O(M)/FS–PS

The name of this genus comes from the Latin *ligua* (a strap), which alludes to the strap-shaped ray florets. The species listed below are, however, quite likely to be found still listed or labelled as senecios in some catalogues and garden centres.

How to grow
Any ordinary soil is suitable, though they do best in moist conditions and associate well with garden pools.

Cut the plants down to ground level in late autumn.

Propagation
Seed sown in warmth (16–18 °C/60–65 °F) in late winter should be ready to prick out into small pots three or four weeks later. Grow on a few degrees cooler, pot on in late spring, and harden off to plant out in the border in early summer. Bigger plants are obtained if flower buds are removed in the first year.

Cuttings of basal growths 5 cm (2 in) long, taken in spring, should root under a cold frame, although they can be tricky. When rooted, pot up and treat as seedlings.

Division is a popular and reliable method. Do it in mid or late spring.

Ligularia przewalskii 'The Rocket'

SOME POPULAR SPECIES	
L. dentata *(Senecio clivorum)* (China) Heart-shaped brownish-green glossy leaves, often flushed purple beneath. Orange-yellow daisy-type flowers 8–10 cm (3–4 in) across, in mid and late autumn. 1.2 m (4 ft). **L. przewalskii** *(Senecio przewalskii)* (China)	Jagged-edged, dark green leaves. Orange-yellow flowers on slender spikes, mid and late summer. 1.5 m (5 ft). **L. stenocephala** *(Senecio stenocephalus)* (China) Jagged foliage. Black-stemmed spikes of bright yellow flowers in mid and late summer. 1.5 m (5 ft).

Lilium
Lily
B/O/FS–PS

An important genus with whole books devoted to these plants. There is space here to mention only half a dozen species and hybrids. But these will be enough to add interest and elegance to any border.

How to grow
Well-prepared, deeply dug soil is necessary if lilies are to do well. Add plenty of garden compost, peat, and grit when planting.

Being such a large group of plants it is not surprising that some of them vary in habit and requirements.

Some species require a lime-free soil, and this has been indicated below where it applies. Others are more amenable to a wide pH range.

Most lilies are basal rooting (roots form from the base of the bulbs), and these are generally best planted in autumn. Stem-rooting lilies, which root from the stem above the bulb as well as from the base, usually need to be planted about two-and-a-half times the height of the bulb.

Few species require staking, but all will benefit from annual mulching.

Lilium Mid-century Hybrid

Lilium 'Fire King'

Propagation

Seed is often slow, and can be difficult. Germination is erratic, and can take from a month to two years. Pre-soak the seeds in tepid water for 24 hours, then sow in pots or trays at least 8 cm (3 in) deep, in early autumn. Bring into warmth (18–21°C/65–70°F) in late winter. Prick out into 15–20 cm (6–8 in) pots during summer. Grow on under a frame for two or three years, uncovering in summer. Finally plant out or pot on in early autumn.

Bulbils (small bulbs) produced in the leaf joints of some species (*L. tigrinum*, for instance) can be removed in summer as the foliage yellows.

Pot them up and then treat as for seedlings.

Offsets (bulblets) provide the easiest method. When bulbs are lifted or repotted in late summer, small bulblets can be removed from the parent bulbs. Plant out or pot up the largest; treat the smaller ones as bulbils.

Bulb scales can be removed from healthy bulbs in early autumn or spring, and set vertically in pots or trays with the tips just showing. Alternatively bury in clear plastic bags half filled with damp horticultural vermiculite. Keep at 10–13 °C (50–55 °F), and pot up singly when growth develops. Then treat as described for bulbils and seeds.

SOME POPULAR SPECIES			
L. auratum (Japan) (Golden-rayed lily) Large, fragrant flowers, perhaps 25 cm (10 in) across. White, spreading petals, striped yellow and spotted purple, late summer and early autumn. There are several varieties. Stem-rooting. Requires lime-free soil. 1.2–2.1 m (4–7 ft). **L. candidum** (Asia Minor) (Madonna lily) Fragrant,	trumpet-shaped white flowers about 8 cm (3 in) long, in early and mid summer. Basal-rooting. Tolerates lime. 1.2–1.5 m (4–5 ft). **L. martagon** (Albania, Eastern Europe) (Turk's-cap lily) Pale or dark purple nodding flowers with reflexed petals and almost black spotting, mid summer. There is also a white form and other varieties. Basal	rooting. Lime-tolerant. 1.2 m (4 ft). **L. regale** (China) Large trumpet flowers, white inside with yellow base, lilac-mauve to purple outside. Fragrant. Stem-rooting. 1.2–1.8 m (4–6 ft) **L. tigrinum** (China, Korea, Japan) (Tiger lily) Orange-scarlet flowers profusely spotted with purple-black. Bulbils form in the leaf axils.	Stem-rooting. Needs lime-free soil. 1–1.8 m (3–6 ft). **L. Mid Century Hybrids** (Garden origin) Flowers, about 10 cm (4 in) wide, are usually borne in dense heads, in early and mid summer. Colours range from yellow through shades of orange to maroon-red. There are several named varieties. Stem-rooting. 1–1.2 m (3–4 ft).

Linum

Perennial flax

H/O/FS

The genus of 230 species includes annuals as well as perennials. *Linon* is an old Greek name used by Theophrastus for some of these plants. It is from plants of this genus that flax and linseed oil are produced.

Linums are generally short-lived, but propagation is easy.

Linum narbonense

How to grow

Ordinary soil in full sun is ideal. The plants are thin and delicate in appearance and are best planted in groups to make an effective display.

Propagation

Seed can be sown in containers under a cold frame or in a greenhouse (18°C/65°F). They should be ready to prick out into small pots after about a month. Harden off, plunge outdoors for the summer, and plant in the border in early autumn.

Cuttings require rather more skill than seed. Use 5 cm (2 in) basal shoots in mid or late spring, and root in sandy compost under a cold frame. Pot up when rooted (usually in about a month), plunge out of doors for the summer, and plant in autumn.

SOME POPULAR SPECIES	
L. flavum (Germany to Russia) Sub-shrubby habit. Narrow, lanceolate leaves. Bright yellow flowers, all summer. 30–45 cm (1–1½ ft). **L. narbonense** (Southern Europe) Dainty clear blue flowers,	about 2.5 cm (1 in) across, on graceful stems well clothed with narrow, pointed leaves, all summer. 45–60 cm (1½–2 ft). **L. perenne** (Europe) Lighter blue flowers than the previous species. 45 cm (1½ ft).

Liriope

Lilyturf

H/AD/FS–PS

A genus of six evergreen hardy perennials, commemorating the nymph Liriope. They are useful because they flower late and are generally tough into the bargain. As the grass-like foliage does not die down in winter, the plant is decorative at all seasons. It is also well able to tolerate drought, the small root nodules or tubers acting as a reservoir.

The hardiness of some species in very severe winters is in doubt, but the most popular species, *L. muscari*, is very hardy.

Liriope muscari

How to grow

The plants are best in light, sandy soil, ideally acid or neutral.

Cut off the spikes after flowering.

Propagation

Division is easy and the normal method, but do not disturb the plants more than once in three to five years. Lift and carefully force clumps apart in early or mid spring. The sections can be replanted immediately, but it is better to pot up young rooted outer pieces in 13–15 cm (5–6 in) pots, and place them under a frame. Plunge outdoors for the summer and plant in early autumn (in cold districts, give them winter protection in a frame and plant out in early or late spring).

SOME POPULAR SPECIES	
L. muscari (Japan, China) Grass-like foliage. Small mauve-blue flowers tightly packed on the spike rather like a tall grape hyacinth	(*Muscari*), early to late autumn. 45 cm (1½ ft). **L. spicata** (China) Similar to the above species but dwarfer and paler. 30 cm (1 ft).

Lobelia

H/O(M)/FS–PS

Lobelias are popular as half-hardy bedding plants, but the border species are very different in appearance. The species listed here make tall, imposing plants to punctuate a bedding scheme, and will enhance a waterside planting, as well as looking good planted as a bold group.

You will, however, have to be prepared to lift *L. cardinalis* and *L. fulgens* for the winter. These two plants are often confused and one may be sold as the other. The plants commemorate Matthias de L'Obel (1538–1616), a Flemish botanist and physician to James I.

How to grow

A rich, moist soil is required for the best results. They grow well in partial shade but will tolerate full sun if not allowed to dry out.

L. cardinalis and *L. fulgens* should be lifted in late autumn and the roots packed in deep boxes of compost and placed in a cold frame or greenhouse for the winter. Separate the pieces of new growth in early spring and pot them up individually to grow on until ready to plant out in late spring or early summer.

Propagation

Seed surface-sown in warmth (18–21°C/65–70°F) in late winter should germinate in two or three weeks. Prick out singly into small pots and grow on a few degrees cooler. Place in a cold frame in late spring, pot on in early summer, harden off and plunge outdoors. Give winter protection, and plant out in late spring.

Cuttings of basal shoots 5–8 cm (2–3 in) long taken in early or mid spring should root at about 16°C (60°F). Pot up when rooted (usually in about a month), and grow on as seedlings.

Division is easy. Lift and pot up mature clumps to overwinter in a cold frame, then in early or mid spring remove rooted young offsets. Pot them up into 10–13 cm (4–5 in) pots, harden off in late spring, and plant out in early summer.

SOME POPULAR SPECIES	
L. cardinalis (Northern USA) (Cardinal flower) Spikes of five-lobed brilliant scarlet flowers, mid to late summer. The true species has oblong to lanceolate green leaves, but hybrids with *L. fulgens* may have	reddish-bronze foliage. Not dependably hardy. 90 cm (3 ft). **L. fulgens** (Mexico) Similar to above, but always purple or reddish-bronze foliage, and taller. Not dependably hardy. 1–1.2 m (3–4 ft).

Lobelia cardinalis

Lupinus
Lupin
H/AD/FS–PS

A genus of 200 species ranging from annuals to sub-shrubs, but the Russell Hybrids derived from *L. polyphyllus* (and other species) are so outstanding that these are the only serious contenders for a place in most borders.

Although relatively short-lived, lupins are easy to propagate yourself.

How to grow
Lupins are not difficult to grow, but they do need deeply-dug soil with plenty of well-rotted manure or compost added.

They do best on acid or neutral soils and do not usually do well on chalky soils.

Cut off dead flower stems.

Propagation
Seed sown in warmth (13–18°C/55–65°F) in late winter will flower the same year. Pre-soak in tepid water for 24 hours before sowing—they should germinate in two to six weeks. Prick out singly, harden off and plant outdoors.

Alternatively, sow in containers under a cold frame in early or mid spring, after pre-soaking. They should be ready to prick out singly into 10 cm (4 in) pots in four to eight weeks. Harden off and plunge outdoors in late spring or early summer. Remove any flower spikes that arise in the first year.

Cuttings are less easy, but will be necessary if you want to propagate a particular plant. Use 5–8 cm (2–3 in) basal shoots in early or mid spring. Root in a sandy cuttings' compost under a cold frame. Pot up into 10 cm (4 in) pots as soon as the leaves start to grow again, usually in about a month. Treat as late-sown seedlings.

Top *Lupinus* 'Harvester'
Bottom *L.* 'Serenade'

SOME POPULAR SPECIES	
L. polyphyllus hybrids (Garden origin) These plants can need little description. The digitate leaves and bold spikes closely packed with	pea-type flowers in early and mid summer are well known. There are named varieties, but mixed seed usually produces good plants. 90 cm (3 ft).

Lychnis
Campion
H/O/FS–PS

A small genus of 12 species of hardy and half-hardy annuals and herbaceous perennials. They include plants a few inches high to over 1.2 m (4 ft), but they are generally bright, colourful plants. Most are good for cutting. The name is derived from the Greek *lychnos* (a lamp), alluding to the brilliantly coloured flowers.

How to grow
Will grow on almost any soil, but are likely to do best on free-draining land. They should be watered freely in summer. Tall species may need the support of twiggy sticks. Mulch in spring, and dead-head if possible.

Propagation
Seed is the easiest and best method for most purposes. Sown in containers in a cold frame in mid spring, the seedlings will usually be ready to prick out in about a month. Harden off then plunge outdoors in late spring or early summer. Plant out in autumn.

L. chalcedonica and *L. coronaria* will usually flower the same year if sown in warmth (13–16°C/55–60°F) in mid or late winter. Prick out singly, harden off, and in late spring plant where they are to flower.

Division can be done in early autumn or in spring, but is sometimes tricky with *L. coronaria* and *L. viscaria*.

Top *Lychnis* x *haageana*
Bottom *L. chalcedonica*

SOME POPULAR SPECIES			
L. x arkwrightii (Garden origin) Green leaves. Brilliant scarlet-orange flowers, all summer. Tends to have a weak habit and is usually short-lived. 30 cm (1 ft). **L. chalcedonica** (Eastern Russia) Light green leafy spikes topped by clusters of scarlet flower heads about 8–10 cm (3–4 in) across,	individual blooms opening to form a cross. Mid and late summer. 90 cm (3 ft). **L. coronaria** (*Agrostemma coronaria*) (Southern Europe) Lanceolate leaves, almost white and covered with downy hairs. Branching spikes of vivid magenta-rose (or sometimes white) flowers. Short-lived. 45–60 cm (1½–2 ft).	**L. flos-jovis** (Europe) Tufty plant with silvery leaves. Bright pink, purple or red flowers in early and mid summer. Short-lived and often treated as biennial. 45–60 cm (1½–2 ft). **L. x haageana** (Garden origin) Green leaves sometimes flushed purple. Orange, scarlet (sometimes pink,	crimson or white) flowers. Short-lived and often treated as biennial. 23–30 cm (9–12 in). **L. viscaria** (Europe, Siberia, Japan) Stems and green foliage are sticky. Purplish flowers early summer, but the form usually grown is the double cerise 'Splendens Plena'. Early summer. 30 cm (1 ft).

Lysimachia

Loosestrife
H/O(M)/FS–PS

A diverse genus of 200 species, those below including a ground-hugger that will thrive in wet soil yet still does well in a dry border, and tall border plants reaching 90 cm (3 ft). All are very easy to grow. Some are inclined to ramble beyond their allotted space.

The name was used by the Greek Dioscorides, probably after Lysimachus, King of Thracia.

How to grow
These plants will adapt to most soils, but they are happiest in moist ground. Tall species may need the support of twiggy sticks.

Propagation
Seed is not widely used, but if obtainable can be sown in containers in a cold frame in early spring.

Cuttings about 8–10 cm (3–4 in) long, taken from non-flowering young shoots of *L. nummularia* root readily in a cold frame if taken in mid spring or early autumn. Pot up when rooted, harden off, and plunge outdoors for the summer. Plant out in autumn.

Division is easy. It is best done in early autumn

Lysimachia clethroides

or early or mid spring. The mat-forming *L. nummularia* can be split up every year if necessary, but do not disturb the others more frequently than every three or four years.

SOME POPULAR SPECIES	
L. clethroides (China, Japan) Lanceolate green leaves, sometimes colouring in autumn. Small white flowers in tapering, arching spikes. 60–90 cm (2–3 ft). **L. ephemerum** (South West Europe) Narrow oblong-lanceolate glaucous foliage. Spikes of	small, star-shaped white flowers with dark centres, mid and late summer. 90 cm (3 ft). **L. nummularia** (Europe) (Creeping Jenny, moneywort) Trailing evergreen with small rounded leaves. 'Aurea' has yellow foliage. Yellow flowers among the foliage. 2.5 cm (1 in).

Lythrum

Purple loosestrife
H/O(M)/FS–PS

A genus of 35 species, taking its name from the Greek *lythron* (black blood) in reference to the colours of the flowers in some species.

Although having the word loosestrife in the common name, the plants are not related to the lysimachias described in the previous entry. They are, however, almost as adaptable—preferring moist ground by the waterside but adapting to the herbaceous border too.

How to grow
Lythrums will grow in moist soils, but will require watering in times of drought, and the plants are happiest in moist conditions.

Feed annually and mulch in spring.

Lift and divide the plants after about three years to maintain vigour.

Lythrum virgatum 'Rose Queen'

Propagation

Seed can be sown under a cold frame or in a greenhouse (16–18°C/60–65°F) in mid spring. Germination will take about a month in the greenhouse, longer in the frame. Prick out into small pots, harden off and plunge outdoors in late spring or early summer. Plant out in autumn.

Cuttings of 6 cm (2½ in) basal shoots taken in mid spring will root in a sandy cutting's compost in a cold frame. Pot up singly when rooted (probably after a month), harden off, and plunge outdoors for the summer. Plant out in early autumn. Division is the easiest method of producing a few plants.

SOME POPULAR SPECIES	
L. salicaria (Northern temperate regions, Australia) A rather coarse plant with green lanceolate leaves and closely-packed spikes of purple-red flowers all summer. It is usually the varieties that are grown, such as 'Robert' (clear pink), 'Dropmore' (purple), 'The Beacon' (rosy-red) and 'Lady	Sackville' (pink). 60–150 cm (2–5 ft). **L. virgatum** (Eastern Europe) Slender, elegant growth. Wiry stems clothed with narrow, pointed leaves. Spikes of rosy-purple flowers. Good varieties are 'Rose Queen' (rose-pink) and 'The Rocket' (deeper pink). 60–90 cm (2–3 ft).

Meconopsis

Poppy
H/O/PS

These are not true poppies (see *Papaver*), but they look like poppies, a fact reflected in their Latin name, which comes from the Greek words *mekon* (a poppy) and *opsis* (like).

The meconopsis described here are perhaps happier in a more natural setting than the herbaceous border. The blue poppy (*M. betonicifolia*) is a woodland plant that always seems to attract an enthusiastic gathering of admirers.

The Welsh poppy (*M. cambrica*) will be at home in the wild garden where it can be aggressive without being a nuisance.

How to grow

M. cambrica will grow in almost any soil and once established should look after itself. Although short-lived it seeds itself freely.

M. betonicifolia is decidedly tricky and likely to disappoint. It may take several years to flower, and then normally it will die. Try removing the first flower buds before they develop, as this is claimed to induce offsets around the crown in some cases, giving the plant a more perennial habit. This species and *M. grandis* prefer a light, well-drained, but rich, soil. Water freely in summer.

Meconopsis cambrica

Propagation

Seed must be sown fresh. Surface sow in containers under a cold frame in late summer. Prick out into small pots in about four to six weeks. Overwinter under the frame, pot on in mid spring, harden off and plunge outdoors in late spring. Plant out in early or mid autumn.

Alternatively, sow in warmth (13–18°C/55–65°F) in early spring. Seedlings have usually appeared within two to four weeks. Prick out singly into small pots. Harden off and treat as described above.

The seedlings can be fickle, and it is worth paying attention to detail when raising plants by this method.

SOME POPULAR SPECIES	
M. betonicifolia (*M. baileyi*) (Tibet, Yunnan, Upper Burma) (Himalayan blue poppy) Large blue (the shade varying from sky-blue to almost purple), about 5–8 cm (2–3 in) across, in early and mid summer.	There is a white form. 1–1.2 m (3–4 ft). **M. cambrica** (Western Europe) Deeply dissected green leaves. Clear yellow, sometimes more orange, single or double flowers. 30 cm (1 ft).

Mimulus

Monkey flower
H/O(M)/FS–PS

A genus of 100 species of hardy annuals and perennials. The perennials described here are generally short-lived but very easily propagated.

The name comes from *mimo* (ape), in allusion to the similarity of the flowers to a monkey's face.

How to grow
Mimulus like a cool, partially shaded position, but they will thrive in full sun if watered freely in dry weather.

Propagation
Seed sown in warmth (18–21°C/65–70°F) in late winter will flower the same year. Surface sow in containers and prick out into small pots. Grow on a few degrees cooler, and harden off before planting out in late spring or early summer.

Sowings made in mid and late summer will flower the following year. Harden off and prick out the seedlings in early and mid summer before plunging outdoors. Give frame protection in winter and plant in mid or late spring.

Cuttings are fairly easy. Make them 5 cm (2 in) long from non-flowering basal growths, and root in a cold frame in mid spring. Pot up singly when rooted, harden off, and plant out in late spring or early summer.

Division of established clumps in early or mid spring is the easiest method.

Mimulus guttatus

SOME POPULAR SPECIES	
M. cardinalis (Oregon to Mexico) Hairy ovate and toothed leaves. Brilliant red flowers with a yellow throat, all summer. 60 cm (2 ft).	**M. guttatus** (North America) Large yellow flowers spotted deep red and brown, all summer. 30 cm (1 ft).
M. cupreus (Chile) Oblong to ovate leaves. Yellow flowers up to 5 cm (2 in) across, maturing to coppery orange and spotted with brown, all summer. 30 cm (1 ft).	**M. luteus** (Alaska to New Mexico) (Monkey musk) Yellow flowers blotched and spotted with red, brown, and maroon, from late spring to late summer. 30–45 cm (1–1½ ft).

Monarda

Bee balm, bergamot
H/O(M)/FS–PS

A genus of 12 hardy annuals and perennials, named after the 16th-century Spanish physician and botanist Nicolas Monardes.

Although in nature they show a preference for wet areas the plants listed here will adapt to the normal herbaceous border quite happily.

Monarda didyma 'Cambridge Scarlet'

How to grow

If possible, grow in moist soil, but do not omit them from the border if you do not have the ideal conditions. Just water in dry weather and mulch the plants well each spring.

After a few years the plants can begin to be invasive, so be prepared to lift, divide, and replant if necessary.

Propagation

Seed can be sown under a cold frame in early or mid spring, or better still in warmth (16–18°C/60–65°F) in late winter or early spring. Prick out into small pots and lower the temperature to 10–15°C (50–58°F). Pot on in late spring, harden off, and plunge outdoors in early summer. Plant out in the border in early autumn.

Division is the usual method, but do not disturb the plants more often than every third year. Do this in early or mid spring.

SOME POPULAR SPECIES	
M. didyma (Eastern USA) Ovate-lanceolate hairy leaves (sometimes dried and used in tea). Whorled	heads of scarlet, pink, purple, or white flowers, all summer. There are several good varieties. 60–90 cm (2–3 ft).

Muscari

Grape hyacinth

B/O/FS

A genus of 60 species of bulbs, taking its name from the Greek *moschos* (musk), some species have a musky scent.

Few small bulbs establish themselves and multiply as well as the grape hyacinths. They are effective as an edging to a bed, or in a bold drift at the front of the border. Either way they will bring early colour while most herbaceous plants are just beginning to stir.

How to grow

Any ordinary well-drained soil is suitable. Plant in early autumn, 5–8 cm (2–3 in) deep and apart.

Clip off dead flower heads.

The bulbs will soon multiply, but leave until they are overcrowded, then lift, separate, and replant.

Propagation

Seed provides a ready means of increase, and the seedlings will normally flower in one to three years in reasonable conditions. Sow in containers in a cold frame between mid summer and early autumn—they should germinate in spring. When the seedlings yellow and dry off in summer (a year after sowing), plant out the small bulbs in their flowering positions in the garden.

Division of clumps is an easy way to increase muscari. Fork up in early summer when the foliage has started to die back. Split up into smaller clumps or separate out individual small bulbs. Small bulbs will have to be grown on for a year before they flower.

SOME POPULAR SPECIES	
M. armeniacum (South Eastern Europe, Western Asia) Grass-like leaves. Densely-packed spikes of small deep blue bells, mid and late spring. There are varieties (such as 'Cantab' and 'Heavenly Blue') in other shades of blue. 25 cm (10 in).	**M. botryoides** (Central Europe) Blue flowers, early to late spring. There is a white variety. 20 cm (8 in). **M. tubergenianum** *(M. aucheri)* (Persia) When in full flower the top half is bright blue, the lower half deep blue. Early spring. 20 cm (8 in).

Muscari tubergenianum

Narcissus
Daffodil
B/O/FS–PS

Top *Narcissus* 'February Gold'
Bottom *N.* 'Llandudno'

Popular bulbous plants that bring much-needed early interest among deciduous shrubs and herbaceous borders. This vast group of plant is divided horticulturally into a dozen groups, each with particular characteristics. Here there is space to mention only a few of the popular kinds that are useful border plants. Any good bulb catalogue will list hundreds of varieties from which to choose.

How to grow
Plant in autumn in large groups so that they form bold clumps or drifts. Keep to one variety in each group unless naturalising in grass.

They will look after themselves, and can be left for years before it is necessary to lift and divide the clumps. Do not be tempted to knot the leaves after flowering—it hardly looks attractive anyway, and can be detrimental to the plant. Just let the leaves die down naturally. They should not be conspicuous among the other herbaceous plants that will be developing.

Propagation
Seed is slow, taking three to six years from sowing to flowering, and other methods provide a more suitable method for amateurs.

Offsets, or small bulbs, removed in summer and replanted immediately will bloom in one to three years.

Chips or bulb segments are the commercial grower's answer to propagation. To try this lift some large bulbs in mid summer, cut each into eight to sixteen segments and dip in a fungicidal solution (use waterproof gloves). Then bury them in moist horticultural vermiculite in plastic bags. Keep at about 21 °C (70°) for 12 weeks, by which time the embryo bulbs that form at the base of the segments should be ready to plant out and grow on for another year.

SOME POPULAR SPECIES	
Daffodils need no description. There are literally thousands of varieties, however, and it is worth consulting a bulb catalogue. For planting among shrubs, or for creating pockets of early interest in the herbaceous	border, it is best to choose from the larger varieties, but those with small cups can be as effective as those with large trumpets. For very early bloom, consider some of the *N. cyclamineus* hybrids such as 'February Gold' and 'Peeping Tom'.

Nepeta

Catmint

H/O/FS–PS

A genus of 250 species, just a few of which are in general cultivation. The catmints are not plants to introduce if wasps and bees worry you, because they always attract insects of this kind when in flower. They are, however, excellent border plants because they combine attractive flowers with pleasing foliage that holds interest long beyond the flowering period. Their compact habit means that they require no staking.

There is much confusion over the naming of the most popular species—*N.* x *faassenii*. The plant is also often sold as *N. mussinii*, but there is in fact a distinct species by that name. Whichever name the plant is sold under, it is worth buying and growing.

How to grow

Ordinary well-drained soil will suit these plants. They will tolerate partial shade, although full sun will produce better plants.

Cut down the plants to just above soil level in autumn.

Propagation

Seed can be sown in containers under a cold frame in late spring and early summer. The seedlings are usually ready to prick off into small pots after three or four weeks. Harden off and plunge outside for the summer. Give winter protection. Harden off and plant out in spring.

Alternatively, sow in warmth (16°C/60°F) in early or mid spring. They should germinate in two to four weeks. Prick out into small pots, harden off in late spring and plunge outdoors. Either plant out in autumn or overwinter under a frame then plant out in spring.

Cuttings are not easy. Use basal growths 5–8 cm (2–3 in) long in April. Use a sandy cutting's compost. Pot up when rooted (usually after five or six weeks), harden off, and plunge outdoors for the summer. Plant out in autumn.

Division of established clumps can be carried out with little trouble in early or mid spring. Use a sharp knife to cut vigorous portions away from the hard centres, which should be discarded.

SOME POPULAR SPECIES	
N. x faassenii (Garden origin) Narrowly ovate grey-green leaves. Whorled spikes of lavender-blue flowers, all summer. Often sold as *N. mussinii*. 30–60 cm (1–2 ft). **N. mussinii** (Caucasus, Persia) Similar to *N.* x *faassenii*	(which is often wrongly sold as *N. mussinii)*, but with leaves more lobed at the base. 45 cm (1½ ft). **M. nervosa** (Kashmir) Bushy habit. Prominently veined lanceolate green leaves. Cylindrical spikes of blue flowers, mid summer to early autumn. 30–60 cm (1–2 ft).

Nepeta 'Six Hills Giant'

Oenothera

Evening primrose

H/O/FS–PS

A genus of about 80 species, the name derived from the Greek words *oinos* (wine) and *thera* (pursuing or imbibing), the roots of one plant being thought to induce a thirst for wine.

These are bold plants, difficult to ignore whether or not you like them. Not all wait until the evening to open their flowers, and some are pleasantly fragrant. There are several good hybrids, most likely to be sold simply under varietal names, such as *O.* 'Fireworks'.

How to grow

Any ordinary, well-drained soil will suit oenotheras, but an open, sunny site produces the best plants.

Dead-head *O. biennis* regularly to prevent self-sown seedlings becoming a nuisance.

Cut the plants down to ground level in late autumn.

Propagation

Seed is the best way to raise these plants. The normal method is to sow in mid spring in a cool greenhouse or cold frame. Prick out singly into small pots when the seedlings are large enough to handle, usually after four or five weeks. Harden off and plunge outdoors in early summer. Plant out in early or mid autumn.

Alternatively sow in warmth (18°C/65°F) in late winter. They should flower the same year. Prick out into small pots and plant out in late spring or early summer, when the risk of frost is past.

Division is fairly easy if done in early or mid spring. This method is not suitable for *O. biennis*, which is a biennial.

SOME POPULAR SPECIES	
O. biennis (North America) Lanceolate leaves. A profusion of pale yellow flowers, up to 5 cm (2 in) across, carried on erect stems, early summer to mid autumn. Self-sown seedlings can become invasive. 1–1.5 m (3–5 ft). **O. fruticosa** (North America) Ovate-lanceolate leaves. Yellow flowers about 2.5 cm (1 in) across all summer. There are improved varieties, such as	'Yellow River'. 30–60 cm (1–2 ft). **O. missouriensis** (USA) Lanceolate leaves on stems that tend to lie flat on the ground. Yellow flowers up to 8 cm (3 in) wide, all summer. They open in the evening. 15 cm (6 in). **O. x tetragona** (Eastern North America) Yellow flowers about 2.5 cm (1 in) across, all summer. 30–90 cm (1–3 ft).

Oenothera missouriensis is a good choice for the front of the border or rock garden

Onopordum
Cotton thistle
H/O/FS–PS

Top *Onopordum arabicum*
Bottom *O. acanthium*

A genus of 40 annuals, biennials and perennials. The name is derived from the Greek words *onos* (an ass) and *perdo* (to eat)—one assumes that donkeys may have been partial to them.

Popularly called the cotton thistle because the stems and leaves are covered with a white tomentum that looks like cotton. Most grow very large. The plants usually die after flowering so are generally treated as annuals.

The species described below are best for the back of a large border or in a wild garden.

The flowers are quite attractive in floral arrangements.

How to grow
Onopordums will grow well in almost any soil, but the richer the soil, the taller the plants are likely to grow.

Remove dead flower heads to prevent self-seeding, otherwise the plants can become a nuisance.

Propagation
Seed is traditionally sown outdoors in late spring, and this works well if soil conditions are good. Sow in shallow drills in a sunny nursery bed, and thin out the seedlings to 30 cm (1 ft) apart, when about 5–8 cm (2–3 in) high. Plant out in their flowering positions in autumn or in mid or late spring.

Alternatively, sow the seeds where they are to flower, placing two or three seeds every 75 cm (2½ ft) where the plants are needed. Thin to one seedling in each position.

Sowing in warmth (16 °C/60 °F) in early spring gives good and more predictable results. Sow two or three seeds to a 9 cm (3½ in) pot and thin to one seedling later. Harden off before planting outdoors in late spring or early summer.

SOME POPULAR SPECIES	
O. acanthium (Great Britain) (Scotch thistle, cotton thistle) Large, spiny leaves covered with thick, cottony tomentum. Tall white stems with prickly heads of violet flowers, mid and late summer. 1.8–2.4 m (6–8 ft).	**O. arabicum** (Southern Europe) Spiny, silvery foliage. Branching, winged stems. Thistle-like purple-red flowers (5 cm/2 in across) produced in mid summer. Hardy biennial. 2.4 m (8 ft).

Paeonia

Peony

H/O(M)/FS–PS

A genus of about 30 species of hardy herbaceous and shrubby perennials, most of them with large and very showy flowers. The flowers are also excellent for cutting.

The genus commemorates Paeon, an ancient Greek physician, said to have been the first to use *P. officinalis* medicinally.

Although a few species have been listed below, it is mainly the garden hybrids that are grown, and these are almost always listed by varietal name alone (*P.* 'Sarah Bernhardt' for instance). If you are thinking of buying peonies, it is well worth obtaining a specialist catalogue where the very many types and varieties are described.

How to grow

Peonies need a deep, rich soil if they are to do well. Incorporate as much compost or manure as possible before planting. Set the crowns just beneath the surface.

Peonies will take a couple of seasons before they settle down into impressive plants. Do not disturb the plants unless really necessary.

Mulch each spring. Water freely and feed with a liquid fertiliser during the growing season. Remove the faded flower heads.

Propagation

Seed can be sown in containers under a cold frame in mid autumn, but germination is often slow. Move the containers of seeds into warmth (16–18°C/60–65°F) in late winter. Prick out the seedlings into 10 cm (4 in) pots, and grow on a few degrees cooler for a month. Move to a cold frame in late spring and plunge outdoors in early summer. Plant out in their final positions in early autumn. They should flower in another three or four years.

Division of mature clumps in early autumn is much easier. Fork up the swollen, fleshy roots and cut into rooted segments, each with two or three good buds. Replant immediately, covering with no more than 3–4 cm (1–1½ in) of fine soil. Plants are not likely to flower for two or three years after being disturbed.

SOME POPULAR SPECIES

P. lactiflora (Siberia, Mongolia)
Deep green, lobed leaves. Flowers about 10 cm (4 in) across. The true species is rarely grown, it is the varieties and hybrids that are cultivated. There are single, double, and semi-double, in shades of pink, red, and white; some are fragrant. Late spring and early summer. 60–75 cm (2–2½ ft).

P. mlokosewitschii (Caucasus)
Deeply lobed leaves topped with bold single yellow flowers 10–13 cm (4–5 in) across, in mid and late spring. 60 cm (2 ft).

P. officinalis (France to Albania)
Large, deeply cut leaves, and large flowers often 13 cm (5 in) across. The true species is seldom grown and it is the various varieties and hybrids that are cultivated. These include single and double forms, mainly in pinks and reds. Late spring and early summer. 75 cm (2½ ft).

Top *Paeonia* 'Bowl of Beauty'
Bottom *P.* 'White Wings'

Papaver

Poppy
H/O/FS

A genus of 100 species, including annuals and biennials as well as herbaceous perennials. They are among the most traditional herbaceous border plants.

The genus takes its name from an ancient Latin name of doubtful origin—possibly derived from the sound made when chewing the seeds.

How to grow

Few border plants can be easier to grow than the poppies. They will grow in almost any soil, preferring one that is deep, poor and dry, but need full sun to look their best.

Staking is usually necessary. Dead-head after flowering.

Propagation

Seed is a fairly easy method. Sow under a cold frame in mid spring, or better still in warmth (13–16°C/55–60°F). Seedlings should be ready to prick out into 10 cm (4 in) pots after about a month. Harden off and plunge outdoors for the summer. Plant out in late summer in their flowering position.

Root cuttings about 5–8 cm (2–3 in) long should root without difficulty if taken in late winter. Insert vertically in pots of sandy cuttings' compost, and pot up when the leaves show. Grow on under the frame until late spring, then harden off and plunge outdoors for the summer. Plant out in late summer.

Division in early or mid spring is straight-forward.

SOME POPULAR SPECIES	
P. orientale (Armenia) (Oriental poppy) Deeply divided, hairy leaves. Large flowers about 10 cm (4 in) across in late spring and summer. Colours include scarlet, pink, and white, usually blotched at the base of the petals. There are several	named varieties which will be found in a catalogue. 60–90 cm (2–3 ft). **P. pilosum** (Asia Minor) Rosette of hairy oblong, toothed leaves. Scarlet or orange flowers about 10 cm (4 in) across, early and mid summer. 60 cm (2 ft).

Top *A form of Papaver orientale*
Bottom *Papaver orientale* 'Mrs Perry'

Penstemon
Beard tongue
H/O/FS

A substantial genus of 250 species, although only a few are good border plants. These are also good cut flowers.

Penstemon 'Large Red'

How to grow
Any well-drained soil will suit penstemons, but they need a sunny position.

The border penstemons are not dependably hardy in bad winters or cold areas, and it is worth cutting them down in autumn and overwintering them under cloches or in a cold frame.

Propagation
Seed is an easy way to produce plenty of plants at minimal cost. Surface sow in warmth (13–18°C/ 55–65°F) in late winter. Prick out the seedlings into small pots and grow on a few degrees cooler. Harden off and plant outdoors in late spring, where they are to flower.

Cuttings are practical for all varieties, but the method is usually used to increase 'special' plants. Use 6–8 cm (2½–3 in) long non-flowering basal shoots in late summer or early autumn. Those taken in late summer should be ready for potting up in early autumn, those taken in early autumn may not be ready to pot up until early spring. Harden off and plant out in mid or late spring.

SOME POPULAR SPECIES	
P. barbatus (Colorado) Lanceolate leaves. Long spikes of pendulous tubular flowers, all summer. Colours include scarlet, pink, carmine, and white. 90–120 cm (3–4 ft).	**P. hartwegii** (Mexico) Lanceolate to ovate leaves. Spikes of bell-type red flowers about 5 cm (2 in) long, in early and mid summer. It is a parent of many modern hybrids. 60 cm (2 ft).

Phlox
H/O/FS

A genus of over 60 species, taking its name from the Greek *phlego* (to burn) or *phlox* (to flame), presumably alluding to the bright flowers. Phlox range from annuals and alpines to large border plants. The two border species have given rise to many fine varieties. They are good for cutting.

How to grow
The key requirements for good border phlox are a sunny position and plenty of moisture during the growing season. They do not do well on dry, sandy soils.

Incorporate plenty of well-rotted compost or manure when planting, and mulch each spring. Water freely in dry weather.

Established clumps will produce better flowers if the weakest shoots are thinned out in spring.

The support of twiggy sticks is usually necessary.

Phlox paniculata

Propagation

Seed is little used as the improved modern varieties do not come true. But if you want to experiment, sow in warmth (16–18°C/60–65°F) in early spring. Germination is usually erratic but prick out into small pots, harden off and plunge outdoors in late spring or early summer. Give winter protection and pot on in mid spring. Harden off and plunge outdoors until planting time in autumn.

Cuttings of 8 cm (3 in) long basal shoots taken in early spring will root in pots of sandy compost under a cold frame. Pot up when rooted, and treat as seedlings. Use root cuttings instead if eelworm is suspected.

Root cuttings will root fairly rapidly in late winter. Make them about 8 cm (3 in) long and keep warm (13–16°C/55–60°F). Pot up when rooted, then treat as seedlings. Cuttings are usually ready to plant out about 20 months after taking them.

Division is easy. The best time is mid autumn or in spring.

SOME POPULAR SPECIES	
P. maculata (Eastern USA) Small, dark green lanceolate leaves, stems with purple spots. Tapering panicles with whorls of purple, pink, or white flowers produced in mid summer to early autumn. 60–90 cm (2–3 ft).	**P. paniculata** *(P. decusata)* (Eastern USA) Lanceolate leaves. Dense panicles of flowers from mid summer to autumn. It is the varieties and hybrids that are grown. Colours include many shades, some with a pronounced 'eye'. 75–90 cm (2½–3 ft).

Physalis

Chinese lantern

H/O/FS

A genus of about 100 species, taking its name from the Greek word *physa* (a bladder), because of the inflated calyx. These unusual plants are grown for their inflated calyces, which look not unlike orange Chinese lanterns. These hang down from arching stems, which can be cut in early autumn and dried.

The nomenclature is slightly confused. The plant often sold as *P. franchetii* is now regarded as a variety of *P. alkekengi*.

How to grow

Will grow well in any ordinary soil. Can be invasive, so be prepared to dig around the plants in autumn to remove underground runners.

Propagation

Seed is normally more successful than division. Sow in warmth (18–21°C/65–70°F) in early spring, barely covering with compost. Germination takes about 21–28 days. Prick out into small pots and grow on at 13–18°C (55–65°F). Harden off and plant out in late spring.

Alternatively wait until mid or late spring and sow in a cold frame or unheated greenhouse. Seedlings are usually ready to prick out within a month. Harden off and move outdoors. Pot on in early summer before plunging the pots.

Division is not difficult. Carefully lift established root clumps in early or mid spring and split into mini-clumps, taking care not to damage the fleshy roots more than necessary.

SOME POPULAR SPECIES	
P. alkekengi (Caucasus to China) Triangular to ovate leaves. White flowers on leafy stems in mid and late summer. Followed by bright orange-red inflated calyces enclosing an	orange-red berry. 30–45 cm (1–1½ ft). *P. a. franchetii* (it may be sold simply as *P. franchetii*) from Japan is more erect, has larger leaves, and fewer but large and deeper red calyces. 60 cm (2 ft).

Fruits of Physalis alkekengi

Physostegia

Obedient plant

H/O/FS–PS

The common name is based on the peculiar flowers, which can be moved right or left on the stem so that the bloom retains that position. It is probably a device of nature to allow the plants to grow on windswept hillsides—by moving, the flowers allow the wind to pass instead of being lacerated in the face of it. Good for cutting.

The Latin name is based on two Greek words: *physa* (a bladder) and *stege* (a covering), a reference to the formation of the calyx.

How to grow

Ordinary soil and sun or partial shade will satisfy these plants.

The stiff, erect spikes need no staking.

Propagation

Seed is a fairly easy and dependable method. Sow in warmth (18°C/65°F) in early spring. Prick out into small pots and grow on a few degrees cooler. Pot on in late spring, then harden off and plunge outdoors in early summer. Plant out in autumn.

Alternatively sow in containers under a cold frame in mid or late spring. Give winter protection and plant out in mid or late spring.

Cuttings normally root without difficulty. Use 6–8 cm (2½–3 in) long basal shoots in mid spring, and insert in a sandy cuttings' compost in

Physostegia virginiana 'Vivid'

a cold frame. Pot up into 10 cm (4 in) pots when rooted. Harden off and plunge outdoors in early summer. Plant out in autumn.

Division in mid autumn or early spring is the easiest method.

SOME POPULAR SPECIES	
P. virginiana (North America) Long, narrow, lanceolate foliage, growing in four columns. Spike of pink,	mauve, or white flowers produced from late summer to mid autumn. 60–90 cm (2–3 ft).

Platycodon

Balloon flower

H/O/FS

The flowers open almost flat, like saucers, and are upward-facing. Before the buds burst open they look like small inflated balloons, hence the common name. The Latin name is derived from two Greek words: *platys* (broad) and *kodon* (a bell), referring to the shape of the flowers.

Platycodon grandiflorus 'Apoyama'

How to grow

Deep rich soil is required for platycodons to do well.

They resent root disturbance, so try to avoid moving the plants once established.

Propagation

Seed is the normal method. Although sowing in an outdoor seed-bed in early spring is often recommended, it is better to sow in containers in a cold frame. Prick out the fragile seedlings as soon as the seed leaves open (usually in about four or five weeks). Delay leads to root damage.

A good alternative is to sow three or four seeds to a pot, subsequently thinning to one plant.

Either way, harden off and plunge outdoors in late spring and early summer. Plant out in mid or late autumn, with the minimum of root disturbance.

Division of well-established plants in early or mid spring is possible but not easy, as the roots are easily broken and are slow to re-establish. Cut the crowns cleanly into segments rather than try to tease or pull pieces apart.

SOME POPULAR SPECIES	
P. grandiflorus (Far East) Blue-green leaves. Blue,	white and pink flowers, late summer. 30–60 cm (1–2 ft).

Polemonium

H/O/FS–PS

Polemonium foliosissimum

A genus of about 50 species of herbaceous perennials with graceful pinnate foliage. The name is taken from the Greek *polemos* (war), possibly alluding to the lance-shaped leaflets.

How to grow

Any reasonable soil will suit polemoniums, but they respond to feeding. Cut down the flower stems once flowering is over.

Propagation

Seed provides a fairly easy means of increase. Sow in containers under a cold frame in early or mid spring, or in warmth (13–16°C/55–60°F). Prick out into small pots and grow on under a cold frame. Harden off in late spring or early summer, and plunge outdoors for the summer. Give winter protection. In mid spring either plant out where they are to flower, or pot on to grow in a plunge bed until autumn.

Division is also easy. Use clumps three or four years old and divide them in early or mid spring.

SOME POPULAR SPECIES	
P. caeruleum (Northern hemisphere) (Jacob's ladder) Pinnate leaves. Spikes of bright blue flowers. There is a white form. Late spring to mid summer. Self-sows prolifically. 60 cm (2 ft). **P. carneum** (Western USA)	Domed mound of pinnate leaves. Profusion of pink flowers, also blue and pale yellow. 45 cm (1½ ft). **P. foliosissimum** (Rocky Mountains) Erect habit. Pinnate leaves. Mauve-blue flowers, all summer. 90 cm (3 ft).

Polygonatum
Solomon's seal
H/O/PS–SH

A genus of 50 herbaceous perennials, taking its name from the Greek *polys* (many) and *gonu* (a small joint), referring to the many-jointed rhizome.

Useful plants for partial or full shade, but perhaps more suitable for the wild or woodland garden than the herbaceous border. A useful cut flower.

How to grow
A moist position is ideal, but *P.* x *hybridum* will also do well in any reasonable soil.

Add plenty of peat or compost when planting, and mulch each spring.

Propagation
Seed is slow to germinate, taking from one to eighteen months. Sow in containers in a cold frame in early or mid autumn. Bring into warmth (13–18°C/55–65°F) in late winter—most seeds should germinate within a month or two. If only a few show, put the remainder back under the cold frame and bring back into warmth again in late winter. Pot up the seedlings into small pots as soon as large enough to handle. Harden off and plunge outdoors in late spring or early summer.

Polygonatum x *hybridum*

Give winter protection, pot on in mid spring, and plunge outdoors again for the summer. Plant out in early autumn, or grow on for another year first.

Division of established clumps of spreading roots is normally successful and easier than raising from seed. Do it in early autumn or early or mid spring.

SOME POPULAR SPECIES	
P. commutatum (USA) (Giant Solomon's seal) Ovate to lanceolate-oblong leaves. Clusters of white flowers about 2.5 cm (1 in) long, in late spring. 1.8 m (6 ft).	**P. x hybridum** *(P. multiflorum)* (Europe) Arching stems, pointed leaves, clusters of small greeny-white flowers in early summer. 60–90 cm (2–3 ft).

Polygonum
Knotweed
H/O/FS–PS

A genus of about 300 species, taking its name from the Greek words *polys* (many) and *gonu* (a small joint), an allusion to the many joints in the stems.

They range from diminutive species for the rock garden to shrubby climbers. Some are rampant plants best avoided (they are not included here), but there are also highly desirable border plants.

How to grow
The species listed here are all adaptable and will grow in any reasonable soil, in sun or partial shade. Cut the old flower stems to ground level in late autumn.

Polygonum affine 'Darjeeling Red'

Propagation

Seed, if obtainable, can be sown in containers in early or mid winter and placed in a vermin-free spot outdoors to chill for a month. Then bring into warmth (13–16°C/55–60°F) to germinate. They are usually ready to prick out into small pots after four to eight weeks. Harden off and plunge outdoors in late spring or early summer, and plant out in autumn.

Division into small clumps, or even into single strands of rhizome (pieces of underground stem) is the normal method. This is an easy method of increasing plants, best done in early autumn or early spring.

SOME POPULAR SPECIES	
P. affine (Nepal) Prostrate growth. Tufts of long, narrow leaves, tinted red in autumn and winter. Erect, long-lasting spikes of pink or red flowers, late summer to mid autumn. Good ground cover. 15–30 cm (6–12 in). **P. amplexicaule** (Himalayas) Leafy, clump-forming plant. Heart-shaped	leaves. Spikes of red flowers, early summer to early autumn. 90 cm (3 ft). **P. bistorta** (Europe) Green foliage becoming bronze in autumn. Pinky-red 'poker' flower spikes in late spring and early summer. A coarse, rather aggressive habit. It is the variety 'Superba' that is usually grown. . 60 cm (2 ft).

Potentilla

Cinquefoil

H/O/FS

Potentillas can be confusing because some are shrubby and others are herbaceous. Both types are worth considering for a mixed border. Only herbaceous species are included here.

How to grow

Potentillas like rich, free-draining soil. Water-logged ground is likely to kill them, but they need plenty of water during the growing season.

Plant in groups rather than individually, as single plants can look rather weedy.

Propagation

Seed is a useful method of raising plants, especially where division is difficult. Sow in containers under a cold frame in early or mid spring. Seedlings are normally ready to prick out into small pots by late spring or early summer. Plant out in the border in early or mid autumn.

Cuttings of 6–8 cm (2½–3 in) long basal growths, taken in mid spring, will usually root in a cold frame. Pot up when rooted (usually after about a month), and harden off. Stand outdoors for the summer. Plant in early autumn.

Division is possible for clump-forming species, although it is not always easy. It is best done in early autumn or early spring.

SOME POPULAR SPECIES	
P. nitida (Southern Europe) Tufted, mat-forming plant with silvery palmate leaves. Pink flowers, mid and late summer. There is a white form. 15 cm (6 in). **P. recta** *(P. warrenii)* (Europe) Digitate leaves. Bright yellow flowers all summer. Tends to be short-lived. 60 cm (2 ft). **P. tabernaemontani** *(P. neumanniana, P. verna)* (Western Europe) (Spring cinquefoil) Carpeting plant with	digitate, evergreen foliage. Bright yellow flowers in mid and late spring. Sometimes a few more flowers later. 5–8 cm (2–3 in). **Garden hybrids** (hybrids of *P. atrosanguinea* and *P. nepalensis*) Coarsely-toothed and lobed grey-green or green strawberry-like leaves. Loose sprays of single or semi-double flowers in shades of yellow, red, orange, and pink, all summer. 30–45 cm (1–1½ ft).

Potentilla nitida 'Rubra'

Primula

H/O(M)/FS–PS

Forms of Primula vulgaris

A genus of about 500 species, the name derived from the Latin *primus* (first). Many species, including the primrose and polyanthus, are spring-flowering.

An invaluable group of plants, despite the small stature of most of them. The majority of species listed here are suitable for grouping in the front of a border, though they will often look more at home in a wild garden or in association with water.

How to grow

Most primulas prefer moist soil and a position in partial shade, although full sun is acceptable if the ground is not too dry.

Incorporate plenty of peat or well-rotted compost at planting time.

Mulch annually, and water freely in very dry weather.

Propagation

Seed-raising is not the slow process sometimes imagined, provided the seed is fresh and stored properly. Use a compost with a high peat content, and never allow the seedlings to dry out completely. Surface sow in warmth (16–18°C/60–65°F) in early spring. Prick out into small pots and grow on at 13°C (55°F). Harden off in early summer and plunge outdoors until early autumn when they can be planted in their flowering positions.

Division is probably easier than seed, and frequently resorted to where there is an existing stock of good plants. Split up into well-rooted rosettes.

Cuttings are not, as a rule, used for border primulas.

SOME POPULAR SPECIES			
P. denticulata (China, Himalayas) (Drumstick primula) Compact rosette of pale green farinose leaves. Dense globular heads about 5 cm (2 in) across of small lilac flowers from early to late spring. There are other colours, including blue, mauve, and white. 30 cm (1 ft).	**P. japonica** (Japan) (Japanese primrose) A 'candelabra' type of primula with spaced whorls of flowers on stems well clear of the oblong-ovate leaves. Magenta-red flowers from late spring to mid summer. There are other shades of red as well as pink varieties and white. 60–75 cm (2–2½ ft).	**P. vulgaris** *(P. acaulis)* (Western and Southern Europe) (Primrose) Rosette of corrugated leaves. Yellow flowers with deeper centres, in early and mid spring. Modern hybrid forms have flowers in a wide range of colours— consult a seed catalogue. 15 cm (6 in).	**Elatior Hybrids** *(P. vulgaris elatior)* (Garden origin) (Polyanthus) Rosettes of basal leaves and heads of flowers carried clear of the foliage. Colours include shades of blue, yellow, red, pink, and white, many with contrasting 'eye'. 23 cm (9 in).

Prunella

Selfheal
H/O/FS–PS

A small genus of seven species, taking its name from the German *Brunella*, based on *Die Braune* (quinsy)—an illness that the plants were supposed to cure.

Useful for filling in gaps where other plants find it difficult to grow. They can be invasive, however, so introduce with care.

Prunella x *webbiana* 'Rosea'

How to grow
Prunelles grow best in an ordinary, fairly moist soil in either partial shade or full sun. Dead head frequently to prevent self seeding.

Propagation
Division is the method normally used, but do not disturb plants more than once every three or four years. Division is best done in autumn or early spring.

SOME POPULAR SPECIES	
P. grandiflora (Europe) Green ovate leaves. Purple flowers in short spikes, mid and late summer. 15–23 cm (6–9 in). **P. x webbiana** (Western Europe) Similar to the previous	species but with broader leaves and more compact spikes. There are several varieties, in shades of pink, and also white. Flowers produced in early and mid summer. 15–30 cm (6–12 in).

Pulmonaria

Lungwort
H/O/PS–SD

A genus of ten hardy herbaceous perennials. The name is derived from the Latin *pulmo* (lung)—possibly because the leaves resemble a diseased lung, or because one species was used as a remedy for lung diseases. Although not spectacular plants, they are useful as a ground cover, especially in shade.

How to grow
Any ordinary garden soil will suit pulmonarias, but they will respond to an annual mulch and to watering in dry spells during the summer.

Propagation
Seed can be used, but the resulting plants are often variable. Sow in a cold frame in mid spring. Harden off, and plunge outdoors in early summer. Plant out in early autumn. Division is the usual method but do not disturb the plants more often than once every five years. It is best done in early autumn or mid spring.

SOME POPULAR SPECIES	
P. angustifolia (Central Europe) (Blue cowslip) Narrow green leaves. Flowers open pink but turn blue, in early and mid spring. 30 cm (1 ft).	**P. officinalis** (Europe) (Spotted dog) Green, narrowly elliptic leaves spotted white. Flowers pinkish-red becoming violet-purple, mid and late spring. 30 cm (1 ft).

Pulmonaria angustifolia

Pyrethrum
H/O/FS

The border pyrethrum was long ago transferred to the genus *Chrysanthemum*, but gardeners and nurserymen almost always refer to these plants as pyrethrums, so that convention has been followed here. The true species of *P. roseum* has long been superseded by hybrids that are much better border plants, and these are the ones usually grown.

The name comes from the Greek *pyr* (fire), probably a reference to the fact that the plants were used medicinally to assuage fevers.

How to grow
Plant in a bright, well-drained soil, in a sunny position. Stake with twiggy sticks, and water freely during the growing season.

Cut back the flowering stems as soon as they are over—a few more flowers may be produced in autumn.

Propagation
Seed sown in warmth (13–16°C/55–60°F) in early spring will normally germinate in five or six weeks. Prick out into small pots, harden off and plunge outdoors where they can be given winter protection. Plant out in their flowering positions in mid spring.

P. 'Eileen May Robinson', *light pink*

Division is an easy and reliable method. Fork up clumps three or four years old and split off or pull away rooted tufts in early spring before new growth begins. Or divide immediately after flowering.

SOME POPULAR SPECIES	
P. roseum *(Chrysanthemum coccineum)* (Caucasus) Bright green feathery foliage. Single or double daisy-like flowers about 5	cm (2 in) across, with a yellow disc. The plants grown are hybrids, colours including shades of pink and red as well as white. 60–90 cm (2–3 ft).

Rodgersia
H/O(M)/FS–SD

A small genus of half a dozen hardy herbaceous perennials, commemorating Admiral John Rodgers, US Navy, (1812–82), commander of an expedition during which *R. podophylla* was discovered. Magnificent foliage plants for the waterside.

Although they flower, it is as foliage plants that they are usually grown.

How to grow
Moist soil and a position sheltered from strong winds are the prime requirements of rodgersias.

Rodgersia pinnata

Enrich the soil with plenty of well-rotted compost or manure at planting time.

Water freely in dry weather.

Propagation

Seed is not generally used to propagate rodgersias, but it can be sown in containers under a cold frame in early spring. They should be ready for planting in their flowering positions two years after sowing.

Division is the accepted and popular method of propagation. Lift mature clumps in early or mid spring, divide the rhizomes, and replant immediately.

SOME POPULAR SPECIES	
R. pinnata (China) Pinnate leaves, bronze-purple in the variety 'Superba'. Branching panicles of small white, pink or reddish flowers, in mid summer. 1–1.2 m (3–4 ft). **R. podophylla** (Japan) Palmate leaves like outspread fingers, starting light green but assuming metallic tints as the season	advances. Panicles of cream to white fluffy flowers, in early and mid summer. 1–1.2 m (3–4 ft). **R. tabularis** (*Astilbe tabularis*) (China) Large green leaves, spreading like an umbrella. Panicles of cream flowers freely produced in mid summer. 90 cm (3 ft).

Rudbeckia

H/O/FS

Rudbeckias are nothing if not bold. Most have a cone-shaped disc surrounded by yellow or orange petals. They are good plants for cutting but also make eye-catching patches of colour in the border.

They commemorate Olaf Rudbeck (1660–1740), a Swedish professor of botany and counsellor of Linnaeus.

How to grow

Rudbeckias are not fastidious about soil, but it should be free-draining. Plenty of well-rotted manure or compost incorporated at planting time will be beneficial. Water freely in dry weather, and apply a liquid fertiliser periodically.

Propagation

Seed provides a useful method of increase—some come true, others are likely to be variable. Sow in containers in warmth (13–18°C/55–65°F) in early spring. The seedlings should be ready to prick out in about a month. Pot on in late spring, harden off, and plunge outdoors for the summer. Plant out in early autumn.

Division is quick, easy, and normally carried out in early or mid autumn or in early spring. Do not disturb the plants more often than once every three years.

SOME POPULAR SPECIES	
R. fulgida (*R. speciosa, R. newmanii*) (USA) (Black-eyed Susan) Bushy habit. Oblong to lanceolate leaves. Yellow or orange daisy-type flowers with a purple-brown central cone. The species is not normally grown, but there are a number of good varieties such as 'Deamii' (yellow flowers 8–10 cm/3–4 in across) and 'Goldsturm' (deep yellow flowers up to 13 cm/5 in across) and *sullivantii*. Mid summer to early autumn. 60–90 cm (2–3 ft).	**R. laciniata** (Canada, USA) Ovate, deeply cut leaves. Yellow flowers, 8–10 cm (3–4 in) across, in late summer and early autumn. There are several varieties, including 'Goldquelle' (deep yellow, double). 1.5–1.8 m (5–6 ft). 'Goldquelle' is about 90 cm (3 ft). **R. nitida** (North America) Broad-petalled single, golden-yellow flowers with a green centre. Late summer and early autumn. 'Herbstsonne' is usually grown. 1.8 m (6 ft).

Rudbeckia fulgida sullivantii

Salvia

Sage
H/O/FS–PS

A large genus of about 700 species. They range from the culinary sage to the bright red bedding plants that most people think of first when salvias are mentioned. The plants below are very different from both types.

The name comes from the Latin *salveo* (save or heal). The name was used by Pliny with reference to its medicinal properties.

How to grow
Any well-drained soil will suit border salvias, but incorporate plenty of well-rotted manure or compost when planting.

S. haematodes will probably need the support of twiggy sticks.

Propagation
Seed is a convenient method of increase, especially for *S. haematodes*, which is not easily propagated by other methods. Sow in warmth (18–21°C/65–70°F) in early spring. Seedlings should emerge in 10–15 days. Prick out into small pots as soon as large enough to handle. Harden off and plunge outdoors in late spring or early summer. Give winter protection, then harden off and plant in their flowering positions in mid or late spring.

Salvia 'May Night'

Division is suitable for increasing some forms, and for those it is a fairly easy method. Lift and split off rooted offshoots of *S. argentea* and replant immediately. If *S. x superba* is well established, this can also be divided successfully.

SOME POPULAR SPECIES	
S. argentea (Mediterranean area) (Silver sage) Grown mainly for its white leaves. The pinkish or purplish-white flowers in mid and late summer are not impressive. Tends to be short-lived. 45 cm (1½ ft). **S. haematodes** *(S. pratensis haematodes)* (Greece) Grey-green ovate, rather corrugated leaves. Erect,	branching habit. Purple flower spikes from early summer to early autumn. 60–90 cm (2–3 ft). **S. x superba** (Garden origin) This plant may be listed as *S. nemorosa*, or *S. sylvestris* of horticulture. Spikes of violet-purple flowers from mid summer to early autumn. 90 cm (3 ft). 'East Friesland' is similar but only 45 cm (1½ ft).

Saxifraga
H/O/PS–SD

A genus of 370 species, the vast majority of which are rock plants But a few are useful for the border.

The name is derived from the Latin words *saxum* (rock or stone) and *frango* (to break)— either because in ancient medicine the plants were used for 'breaking' stones in the bladder, or because of a supposed ability of the roots to penetrate and assist the breakdown of rocks.

Saxifraga umbrosa

How to grow

The species listed here all do well in ordinary soil, in a position in partial or full shade. *S. oppositifolia*, a rock plant, does particularly well on alkaline soils.

Propagation

Seed is best sown in gritty compost under a cold frame in mid or late spring. Prick out singly into small pots when the seedlings reach the four-leaf stage, usually six to ten weeks after sowing. Overwinter under the frame, harden off and plunge outdoors in late spring. Plant out in early autumn.

Cuttings of non-flowering tip growths of *S. oppositifolia* root readily in a cold frame in late spring or early summer. They should be ready to pot up in late summer. Keep under the frame until late spring, then harden off and plunge outdoors until planting time in early autumn.

Division is the easiest method, best done after flowering.

SOME POPULAR SPECIES	
S. fortunei (China, Japan) Orbicular leaves with lobed edges. Green above, reddish beneath. Foam-like sprays of small white flowers in mid and late autumn. 30–45 cm (1–1½ ft). **S. oppositifolia** (Northern Asia, North America, Europe) Mat-forming plant studded with small rose-pink flowers in early and	mid spring. Other shades include red, purple, and white. 2.5 cm (1 in). **S. x urbium** (*S. umbrosa*) (London pride) Actually a hybrid between *S. umbrosa* and *S. spathularis*, but resembles *S. umbrosa* and may be sold under that name. Rosettes of thick, slightly fleshy evergreen leaves. Panicles of star-shaped pink flowers in late spring. 30 cm (1 ft).

Scabiosa

Pincushion flower
H/AK/FS–PS

A genus of 100 species of annuals and herbaceous perennials. Traditional border plants, but also worth growing just for a supply of cut flowers.

The genus takes its name from the Latin *scabies* (the itch) a disease that the plants were said to cure.

Scabiosa caucasica

How to grow

A fertile, well-drained soil is necessary for good results. They are best on neutral or alkaline soil.

Cut the stems down to ground level in late autumn.

Propagation

Seed is fairly easy. Sow in warmth (13–16°C/55–60°F) in early spring. The seedlings should be ready to prick out into small pots after four to eight weeks. Harden off and plant outdoors.

Cuttings can be a bit tricky. Use 5 cm (2 in) basal shoots in early spring, and root in containers of sandy cuttings' compost under a cold frame. Pot up when rooted (usually in four to six weeks). Harden off and plunge outdoors for the summer, then plant out in early autumn.

Division of large clumps in early or mid spring is the easiest method of propagation.

SOME POPULAR SPECIES	
S. caucasica (Caucasus) Low-growing foliage topped with mauve 'pincushion' flowers on long stems, mid summer to mid autumn. 'Clive Greaves' is a superior mid-blue variety. Other	colours include blue and white. 60 cm (2 ft). **S. graminifolia** (Southern Europe) Narrow foliage. Pinkish-mauve flowers from early summer to early autumn. 25 cm (10 in).

Schizostylis
Kaffir lily
H/O(M)/FS

Schizostylis coccinea 'Mrs Hegarty'

A genus of two rhizomatous perennials, the name derived from the Greek *schizo* (to split) and *stylos* (a style), reflecting the deeply divided style.

These plants are particularly useful because they flower late—even at the approach of winter. They are also good for cutting, though the chances are you will not want to deplete the border of colour at a time when little else is in flower.

How to grow
A moist, fertile soil is ideal, though they will grow in ordinary soil if watered well in dry weather.

The plant is not reliably hardy in cold areas, and it may be necessary to protect the roots with a thick layer of bracken, peat or leaves in winter, held in place with netting or something similar.

Propagation
Seed is slow and little used. If you want to try, sow in warmth (13–18°C/55–65°F) in early spring. They will be ready to plant in their flowering positions two years from sowing (give winter protection in the meantime).

Division is moderately easy and the usual method. Lift in early or mid spring and split up into small clusters of offsets before replanting.

SOME POPULAR SPECIES	
S. coccinea (South Africa) Narrow, sword-like leaves. Loose spikes of red or pink flowers, rather like	small gladioli but more star-shaped, on wiry stems. Mid and late autumn. 60–90 cm (2–3 ft).

Sedum
Stonecrop
H/O/FS

A large genus of 600 species, most of them rock garden plants, but a few are larger and belong in the herbaceous border. *S. spectabile* and its varieties are especially useful because they flower after most of the other border plants have finished.

Sedum roseum

How to grow
Sedums are tolerant plants and will grow well in most soils. They even do very well on sandy, gravelly, and chalky soils. *S. telephium*, however, does best in a moisture-retentive, but free-draining soil.

Propagation
Seed is sown in early spring, ideally maintaining 10°C (50°F). Prick out the seedlings about 4–5 cm (1½–2 in) apart in trays of five or six to a 13 cm (5 in) half-pot. Pot up singly into 8 cm (3 in) pots before the leaves of one plant touch those of the next. Harden off and plunge outdoors for the summer, and plant out in early autumn.

Cuttings of 3–8 cm (1–3 in) long non-flowering stems can be rooted under a cold frame from mid spring to mid summer. Pot up singly into 9 cm (3½ in) pots when rooted (usually after four or five weeks). Harden off and plunge outdoors. Give winter protection, and plant out in mid spring.

Division is the easiest method. After cutting down the old stems in autumn, dig up the roots of well-established plants, divide them and replant without delay. Alternatively divide them in early or mid spring as described, or use small fleshy segments (but you will need to pot these up and grow under a cold frame for a year).

SOME POPULAR SPECIES	
S. roseum *(S. rhodiola)* (Cold and temperate areas of the Northern Hemisphere) (Rose root sedum) Neat habit. Glaucous foliage. Showy heads of pale yellow flowers in mid and late spring. Roots rose-scented when dry. 25 cm (10 in). **S. spectabile** (China) (Ice plant) Fleshy pale grey leaves and stems. Wide	heads of pink or red flowers from late summer to mid autumn. 30–60 cm (1–2 ft). **S. telephium maximum** *(S. maximum)* (Europe) The form usually grown is 'Atropurpureum', which has deep purple broadly obovate leaves and stems. Pink flower up to 15 cm (6 in) across in early and mid autumn. 45 cm (1½ ft).

Sidalcea
H/O(M)/FS–PS

A genus of 25 species of hardy herbaceous perennials, the name a composite of two related genera: *Sida* and *Alcea*.

One of its old common names, seldom if ever used now, was miniature hollyhock, and this is fairly descriptive of the flower.

How to grow
Any ordinary garden soil will be suitable, but though the plant will tolerate some shade, full sun is better.

Tall varieties may need staking.

Propagation
Seed sown in warmth (16–18°C/60–65°F) in early spring, is normally ready to prick out into small pots within four to six weeks. Grow on a few degrees cooler. Harden off and plunge outdoors for the summer. Plant out in early or mid autumn.

Alternatively, sow in containers under a cold frame in mid spring. Prick out and harden off, then plunge outdoors for the summer. Over-winter under the frame, then plant out finally in mid spring.

Division is a popular and easy method. It is best done in early or mid spring.

SOME POPULAR SPECIES	
S. malviflora (Western USA) Basal leaves rounded and lobed, stem leaves narrower and divided.	Spikes of mallow-like flowers in mid and early autumn. There are several varieties in shades of pink. 75–90 cm (2½–3 ft).

Sidalcea malviflora

Solidago
Golden rod
H/O/FS–PS

A genus of 100 species, the name derived from the Latin *solido* (to make whole, heal), alluding to its healing properties.

Older gardeners will remember when the golden rods were coarse, ungainly plants but modern varieties are worth a place in any border.

How to grow
These easy-to-grow plants will thrive in almost any soil, but tall varieties may require staking.

Cut back the stems in autumn.

Propagation
Seed is fairly easy to germinate and if sown in warmth (16–18°C/60–65°F) in early spring, should be ready for pricking out into small pots in three or four weeks. Harden off in late spring, and either plant directly into their flowering

Solidago 'Golden Spire'

positions or pot on and plunge outdoors until planting time in autumn. Division is more popular, and is easier and quicker than seed. Cut down old flowering stems before forking up large clumps after flowering.

SOME POPULAR SPECIES	
S. x hybrida (Garden origin) These are the garden hybrids usually sold under varietal names, such as 'Goldenmosa' and 'Mimosa'. All have	terminal plumes or sprays of small yellow flowers between mid summer and mid autumn. Consult a specialist catalogue for varieties. Heights range from 30–150 cm (1–5 ft).

Stachys
Lamb's ears, lamb's tongue
H/O(D)/FS–PS

A genus of 300 species. Those described here are useful plants as ground cover for the front of the border. The name comes from the Greek *stachus* (a spike), alluding to the pointed inflorescence. *S. byzantina* is perhaps more often sold under its own name of *S. lanata.*

How to grow
Successful in most soils, and will even do well on a dry site. Trim the plants back in late autumn.

Propagation
Seed is sometimes used by commercial growers but it is often fickle. Sow in warmth (16–21°C/60–70°F) in mid summer. Germination is likely to take about a month. Prick out singly into small pots, and grow on at 13–18°C (55–65°F) for

Stachys lanata (right) *and S. macrantha*

about three weeks. Then move out under a cold frame, where the plants should be left to over-winter. Harden off and plant out in early or mid spring. Division is easier. Early autumn and early and mid spring are good times.

SOME POPULAR SPECIES	
S. byzantina *(S. lanata)* (Caucasus to Iran) Mat-forming. Densely hairy. Mauve-pink flowers, mid-summer. 30–45 cm (1–1½ ft).	**S. macrantha** (Caucasus) Mat-forming. Hairy mid green leaves. Whorls of purple, rose-purple, or violet flowers from late spring to mid summer.

Stipa
Feather grass
H/O(D)/FS

A genus of 300 species of annual and perennial grasses, the perennial here being useful for a dry site.

Apart from their role in the border, they also make good specimen plants in a lawn bed.

The flowers can be cut in mid summer and dried for winter use.

How to grow
A light, fertile soil is best, but these plants will grow in most soils.

Propagation
Seed is a sensible method if you want to start off a collection or need a large number of plants. Sow in warmth (13–16 °C/55–60 °F) in late winter or early spring. Prick out three or four seedlings to a 13 cm (5 in) pot. Harden off and plant outdoors in late spring or early summer.

Stipa gigantea

Where seed-bed conditions are good, seed can be sown outdoors and the seedlings transplanted to their flowering positions in the first weeks of early summer, but this is an unreliable method.

Division of established clumps in early or mid spring is easy.

SOME POPULAR SPECIES	
S. barbata (Asia Minor) Compact tufts of grass. Slender arching flowering stems with 30 cm (1 ft) grass plumes at the ends, all summer. 60 cm (2 ft).	**S. gigantea** (Spain) Dense clumps of leaves. Silvery flower plumes tinged purple, early and mid summer. 1–1.2 m (3–4 ft).

Stokesia
Stokes aster
H/O/FS

Stokesia laevis

A genus of a single species, commemorating Jonathan Stokes (1755–1831), an English doctor and botanist.

How to grow
A light, well-drained soil suits this plant best. Cut down top growth in late autumn.

Propagation
Division is the only practical method of increase. Do this in mid spring. In cold areas, pot up the divisions into 15–20 cm (6–8 in) pots and place under a cold frame. Harden off and plant out in early summer.

SOME POPULAR SPECIES	
S. laevis (South Eastern USA) Flowers about 5–8 cm (2–3 in) across, with deeply notched florets,	from late summer to mid autumn. Lavender-blue but also white, blue, lilac, or purple. 30–45 cm (1–1½ ft).

Thalictrum

Meadow rue
H/O/FS–PS

A genus of 150 species of hardy herbaceous perennials. The plants below are worth growing just for their divided, dainty foliage, which resembles that of a maidenhair fern, but there is the bonus of flowers too. Both flowers and foliage are attractive for cutting.

How to grow
Will grow well in most soils, but they respond to feeding. They enjoy sunshine but will thrive in partial shade.

Mulch each spring. Staking is usually unnecessary.

Propagation
Seed provides an easy way to propagate these plants. Sow in warmth (13–16°C/55–60°F) in early spring. The seedlings should be ready to prick out into small pots after three or four weeks. Then lower the temperature by a few degrees, and pot on in late spring. Harden off and plunge outdoors for the summer. Give winter protection and plant out in mid or late spring.

Seed can also be sown in pots under a cold frame in mid spring. Seedlings should be ready to prick out into 10 cm (4 in) pots about five weeks later. Harden off and plunge outdoors for the summer. Give winter protection and plant out in mid or late spring.

Division is an easy, reliable method. It is best done in early or mid spring.

Thalictrum aquilegifolium

SOME POPULAR SPECIES	
T. aquilegifolium (Europe, Northern Asia) Pinnate grey-blue leaves, resembling those of an aquilegia. Panicles of fluffy mauve or purple flowers in early summer. There are white and purple varieties. 90 cm (3 ft). **T. dipterocarpum** (Western China) Green, slightly glaucous leaves. Pyramidal panicles of small mauve flowers with conspicuous yellow anthers, all summer. There is a white variety, and 'Hewitt's Double' has	double mauve flowers. 1.2–1.5 m (4–5 ft). **T. flavum** (Europe) (Yellow meadow rue) Deep green, almost fern-like, divided leaves. 15 cm (6 in) heads fluffy yellow flowers in mid and late summer. *T. f. glaucum* has grey-green leaves. 1.2–1.5 m (4–5 ft). **T. minus** (*T. adiantifolium*) (Europe) Grey-green maidenhair-type leaves. The yellowish-green flowers in mid summer are insignificant. 30–90 cm (1–3 ft).

Tiarella

H/O(M)/PS–SD

A genus of seven species, taking its name from the Greek *tiara* (a turban). The name is literally little turban, referring to the shape of the seed pods.

Tiarella cordifolia

The common name of foam flower (applied mainly to *T. cordifolia*) reflects the airy fuzziness of the heads of white flowers.

Useful as ground cover in moist shade, but they also have a place in the border if planted in bold drifts.

How to grow
A moist, peaty soil is ideal. Dry soil is a common cause of failure.

Propagation
Seed sown in containers under a cold frame in early or mid spring is usually ready for pricking out in about a month. Plant five or six seedlings to a 13 cm (5 in) pot. Pot up singly in small pots in early summer, and grow on under a shaded frame. Overwinter in the frame and plant out in mid or late spring.

Division is easy, except for *T. wherryi*, which is better raised from seed. Division is best done in early or mid autumn, or mid or late spring.

SOME POPULAR SPECIES	
T. cordifolia (North America) (Foam flower) Scalloped, heart-shaped leaves, taking on ruddy tints in autumn and winter. White fluffy flower spikes, late spring and early summer. Excellent ground cover. 23 cm (9 in). **T. polyphylla** (Himalayas, China, Japan) Clump-forming plant with lobed leaves. Sprays of	ivory-white flowers, early and mid summer. 30 cm (1 ft). **T. wherryi** (South Eastern USA) Tufty plant, less spreading than most species. Ivy-shaped leaves turn reddish in autumn and winter. Feathery spikes of creamy-white flowers tinged pink, produced throughout the summer. 30 cm (1 ft).

Tradescantia

Spiderwort, trinity flower
H/O(M)/FS–PS

The tradescantias that most people know are the foliage houseplant type, but there are some very different hardy border plants. These are not everyone's idea of a desirable border plant— there is rather a lot of leaf in relation to flower— but they are easy to grow.

The plants normally grown are hybrids of *T. x andersoniana* and *T. virginiana*, and are likely to be listed simply as Tradescantia followed by a varietal name.

The genus commemorates John Tradescant, gardener to Charles I.

How to grow
Any ordinary, well-drained soil is suitable.

Staking with twiggy sticks may be necessary, and slugs are often a problem.

Propagation
Seed can be sown in early spring, in containers under a cold frame, or better still in warmth (13–18°C/55–65°F). Harden off before plunging outdoors for the summer. Plant out in early autumn.

Division is the normal practice, and is best done in early or mid spring as the plants are starting into growth. A quick and easy method.

SOME POPULAR SPECIES	
Garden hybrids Almost all the tradescantias grown in gardens are hybrids of *T. x andersoniana* or *T. virginiana*. They have dull green strap-shaped leaves and clusters of three-	petalled flowers about 2.5 cm (1 in) across nestling among the leaves. The flowers appear throughout the summer, in shades of blue, purple, rose, and white. 60 cm (2 ft).

Tradescantia x *andersoniana*

Trillium
Wake robin, wood-lily
H/O/PS

A genus of 30 species of hardy rhizomatous herbaceous perennials, taking its name from the Latin word *triplum* (triple), alluding to the three-parted leaves and flowers.

Trillium grandiflorum

These are very distinctive plants: leaves, petals, sepals are all in groups of three. They are not at home in the normal herbaceous border and should really be grown in a woodland setting.

How to grow
Plant the rhizomes in groups, 8–10 cm (3–4 in) deep, in autumn. A fibrous, leafy soil is ideal for these plants. Moisture is important but the soil should never become waterlogged.

Do not disturb once the plants are established.

Propagation
Seed is very slow to germinate, usually taking from 18 months to three years.

Division is a more practical method for amateurs, but trilliums resent root disturbance and flowering is likely to be delayed for a year or two. Lift large clumps in late summer and divide the rhizomatous roots into smaller clumps.

SOME POPULAR SPECIES	
T. grandiflorum (Eastern USA) Pure white flowers, the petals almost triangular in outline, becoming pinkish with age. Set off against a whorl of leaves behind the flower. There is a pink variety, and a double. 30–45 cm (1–1½ ft).	**T. sessile** (Central USA) Leaves marbled grey and green. Red to maroon flowers. 15–30 cm (6–12 in). **T. undulatum** (Canada, USA) White flowers zoned purple at the base. 30 cm (1 ft).

Trollius
Globe flower
H/O(M)/FS–PS

A genus of 25 hardy herbaceous perennials associated with waterside planting, but they also make good border plants if given partial shade.

How to grow
Trollius will grow well in sun or shade if given moist soil, but in a normal border they need partial shade to do well. Water freely in dry weather. Cutting the stems back after flowering sometimes encourages a few more flowers too.

Propagation
Seed is traditionally sown in containers in autumn under a cold frame—but the seedlings may not appear for a year or more.

Trollius x *hybridus* 'The Globe'

Commercially, the modern practice is to buy in pre-treated seeds ready for sowing. At the time of writing these are not available to amateurs, but may be in the future. In the meantime, try pre-chilling them in the fridge before sowing in warmth (16–21°C/60–70°F) in mid spring. Prick out into 10 cm (4 in) pots, harden off and plunge outdoors for the summer. Plant out in early autumn, or give winter protection and plant in mid spring.

Division is easier than seed. In early autumn or early or mid spring, split up the root into a few sizeable offsets with fibrous roots.

SOME POPULAR SPECIES	
T. x hybridus *(T. europaeus)* (Garden origin) Lobed leaves topped with buttercup-like yellow or orange flowers from late spring to mid summer. Most of the varieties listed as *T. europaeus*, actually a distinct species, are probably hybrids. 45–60 cm (1½–2 ft).	**T. ledebourii** (Eastern Siberia) Finely cut foliage. Deep yellow or orange flowers with wide-spreading sepals, the petals standing more erect. Prominent stamens give the centre of the flower a well-filled appearance. Early and mid summer. 75 cm (2½ ft).

Verbascum
Mullein
H/O/FS–PS

A genus of 300 species of hardy biennial and perennial herbaceous plants and sub-shrubs. The leaves are usually large and white-felted, and there is the bonus of attractive flowers too.

Some of the perennial species seed too freely and can become invasive, but this is easily resolved by cutting off faded flower spikes.

How to grow
Ordinary well-drained soil is suitable for verbascums. Full sun is better than partial shade, but the plants will grow well in either.

Tall plants may require staking.

Dead-head to prevent self-seeding. Cut the plants back to ground level in late autumn.

Propagation
Seed can be sown outdoors in drills in mid or late spring if conditions are suitable. Thin to 15–20 cm (6–8 in) apart. Plant out in autumn. This is often a disappointing method.

Better results can be obtained by sowing in containers in a cold frame in mid spring or mid summer. Prick out the seedlings into 13 cm (5 in) pots, harden off and plunge outdoors. Mid spring-sown seedlings should be ready to plant out in early autumn. Protect those sown in mid summer, and plant out in mid spring.

Root cuttings are easy. Make them about 8 cm (3 in) long and set them vertically in containers under a cold frame in late winter. Pot up into 13 cm (5 in) pots about mid spring, when at the three-leaf or four-leaf stage. Harden off and plunge outdoors for the summer. Plant out in early autumn.

Division can be done in early or mid spring. Although not difficult the plants are often inferior to those grown from root cuttings. Lift the roots, cut into segments, making sure each has one or more good buds, and trim back long, straggly roots.

SOME POPULAR SPECIES	
V. chaixii *(V. vernale)* (Central and Southern Europe) Grey-green leaves covered with white tomentum. The type has yellow flowers with a purple eye. There is a white form. Mid and late summer. 90 cm (3 ft).	**V. x hybridum** (Garden origin) Most of these are hybrids of *V. phoeniceum*. Usually offered under varietal name alone, such as 'Gainsborough' (grey leaves, yellow flowers), 'Pink Domino'. Summer. 1–1.8 m (3–6 ft).

Verbascum 'Pink Domino'

Verbena

H/O/FS

A genus of 250 species of annuals and herbaceous perennials. The name is an ancient Latin one used for *V. officinalis*.

The species here can be grown as perennials in favourable areas, but it is best to treat them all as annuals, as they will flower the first year from seed.

How to grow
These verbenas require a well-drained soil enriched with plenty of well-rotted compost or manure at planting time.

Best treated as annuals, otherwise lift the plants and overwinter under cover.

Propagation
Seed is usually the most convenient method. Sow in warmth (18–21°C/65–70°F) in mid or late winter. Barely cover the seed with compost, but germinate in darkness. Prick out the seedlings (germination usually takes three or four weeks), and grow on at 13–18°C (55–65°F). Harden off and plant out in flowering positions in late spring or early summer.

Cuttings can be used. Overwinter the stools (roots) under cover and move into warmth (13–16°C/55–60°F) in late winter. In early spring take 8 cm (3 in) young, preferably rooted, shoots as cuttings. Pot up singly and treat as seed-raised plants.

Verbena rigida

Division is easy, but the plants are usually inferior to those raised from seeds or cuttings. Overwinter and start off the stools as for cuttings. Then in mid spring split them up, potting up the rooted divisions into 10–13 cm (4–5 in) pots, making sure each has at least two good shoots.

SOME POPULAR SPECIES	
V. bonariensis (South America) Oblong, mid green leaves. Erect, stiffly branching, vigorous growth. Heads of purplish-lilac flowers produced throughout the summer. 1.2 m (4 ft).	**V. rigida** *(V. venosa)* (Brazil, Argentina) Compact, bushy habit. Ovate, dark green leaves. Clusters of small claret-purple or violet flowers from mid summer to mid autumn. 30–60 cm (1–2 ft).

Veronica

H/O/FS–PS

A genus of 300 species of hardy and half-hardy annual and perennial herbaceous plants. Older gardeners may still think of some of the 'shrubby veronicas' under this name, but these now have their own genus: *Hebe*. The plants listed here are the herbaceous veronicas, among them some very desirable border plants.

Veronica teucrium 'Trehane'

How to grow

Although veronicas will grow in most soils they do best in those enriched with plenty of well-rotted compost or manure.

They will grow well in partial shade, but are likely to do even better planted in a position in full sun.

Propagation

Seed can be sown in warmth (16–20°C/60–68°F) in early spring. The seedlings should be ready to prick out into small pots in about four to eight weeks. Harden off and plunge outdoors for the summer. Either plant out in early autumn, or give winter protection, harden off and set out in spring.

Division is easier. This is best done in early or mid spring.

SOME POPULAR SPECIES	
V. gentianoides (Caucasus) Tufts of foliage resembling gentians. Porcelain blue flowers from mid spring to early summer. 30–60 cm (1–2 ft). **V. incana** (Caucasus) Silvery leaves and stems. Deep blue flowers in mid and late summer. 30 cm (1 ft). **V. longifolia** (Central Europe, Northern Asia) Long, lanceolate, toothed leaves. Long spikes of bright blue flowers	produced in summer. 1–1.2 m (3–4 ft). **V. spicata** (Europe) Dark, deeply veined leaves. Small bright blue flowers in short spikes in mid and late summer. There are white and pink varieties. **V. teucrium** (Southern Europe, Northern Asia) Spreading, low-growing habit. Small-leaved slender stems with blue flowers from late spring to late summer. A good carpeter among bulbs. 20–30 cm (8–12 in).

Zauschneria

Californian fuchsia

H/O/FS

A genus of four half-hardy shrubby perennials commemorating J. P. J. Zauschner (1737–99), professor of natural history at Prague.

Zauschneria californica 'Glasnevin'

How to grow

A suitable plant for rather poor, sandy loam and a hot, sunny position. Unfortunately winter losses are a distinct possibility so it is worth propagating a few plants each year to have on hand as replacements.

Cover the plants with bracken or leaves, held in place with netting, in cold districts. Although not dependably hardy, they are worth persevering with because of the contribution they make to late colour in the border.

Cut the plants down to just above ground level in early spring.

Propagation

Seed can be sown in warmth in early spring. Prick off into small pots, and plant outdoors in late spring or early summer.

Cuttings of non-flowering shoots can be rooted in a sandy compost in early autumn and overwintered in a greenhouse.

Division in spring is possible but not easy. It is best not to depend on this method of propagation in case the parent plant is killed over the winter.

SOME POPULAR SPECIES	
Z. californica Bushy habit. Stems densely clothed with narrow, grey-green leaves.	Bright scarlet flowers in loose spikes from late summer to mid autumn. 30–45 cm (1–1½ ft).

Terms & Techniques

This section includes an explanation of technical and horticultural terms used in the first part of the book, as well as some that you might encounter in other books on garden flowers.

It is more than just a glossary, however. Where relevant, the practical implications are explained, and for practical tasks the jobs may be illustrated with step by step illustrations.

We have not included specific chemicals for pest and disease control because these vary from country to country, and new ones are continually appearing, sometimes to replace current controls.

If you familiarise yourself with this part of the book you should find it a useful aid to propagating and caring for border plants, and a source of explanation for those terms that you may not know.

Acid compost/soil

A compost or soil with a low pH (*see* pH), necessary for those plants affected by too much lime. *See* Chlorosis, Compost, and Ericaceous compost.

Alpine

A term used loosely to describe 'rock' plants—low-growing plants suitable for a rock garden. Many are actual mountain plants that qualify for the technical use of the term, which describes those plants native to the alpine zone between the tree line (upper limit of tree growth) and the permanent snow line. Generally the term is used more loosely to include other dwarf or low-growing plants.

In this book only the more rampant 'alpines', such as aubrietas and *Alyssum saxatile* have been included as being suitable for the herbaceous or mixed border.

Alternate

Leaves that arise first on one side of the stem and then on the other.

Annual

A plant that germinates, grows, flowers, then dies within a season. Usually they are sown and die within the same year, but some hardy annuals can be sown in the autumn to overwinter and flower earlier than normal the following year. This is a kind of biennial treatment. Good varieties for this treatment include calendula, cornflower, godetia, larkspur, nigella and annual scabious.

Annual flowers have not been included in this book, although they can be very useful for filling in gaps in a border while the perennials are becoming established.

Aphids

A group of insects of which greenfly and blackfly are the most common. These can attack a wide range of flowers as well as fruit and vegetables.

Fortunately they are easily controlled by a wide range of insecticides.

Aster wilt

A form of wilt that affects species of aster—the Michaelmas daisy, *A. novi-belgii* being particularly vulnerable, although *A. novae-angliae* is more resistant.

The disease, caused by a fungus in the roots, results in the plants wilting and collapsing. Destroy affected plants, and do not replant asters in that position as the disease is soil-borne.

Balled plant

A term used in connection with trees and shrubs, meaning a plant grown in a field and lifted with a ball of soil that is then wrapped in hessian or a plastic material to keep the soil around the roots. Conifers and evergreens such as hollies are sometimes sold this way. *See also* Bare root plant, Container-grown plant, and Pot-grown plant.

Bare-root plant

A plant lifted in the dormant season without a ball of soil. Bare-root plants are now much less common than they used to be, but will transplant successfully if planted at the right time. Soak the roots before planting if they are very dry.

Bare-root plants are most likely to be received by post, as plants removed from their pots are generally easier to pack for delivery. In the case of herbaceous plants, the roots or root-ball may be wrapped in damp newspaper or moss. Plants bought at garden centres are nearly always pot-grown.

Basal shoot

A shoot arising from the crown (base) of the plant. Basal shoots are sometimes used as soft stem cuttings and these are generally taken in spring as the plant is beginning to grow again.

Beard

A tuft or row of (usually long) hairs. Bearded irises have a hairy patch along the fall petals.

Bedding plant

A plant used for seasonal planting—usually for a summer display (summer bedding), but sometimes for a spring display (spring bedding).

Although half-hardy annuals are commonly used for summer bedding, hardy perennials, and of course bulbs, can also be used. The plants are lifted when the display is over and either discarded or, in the case of suitable perennials, saved for the following year.

Bicolor

Two-coloured. Usually applied to flowers in which one colour is contrasted with another. *Gaillardia* 'Mandarin' is illustrated.

Biennial

A plant sown one year to flower the next, after which it dies. Some perennials are best treated as biennials (hollyhocks for instance), and certain varieties of some biennials (such as foxgloves) can be treated as annuals by starting them off early in a greenhouse.

Bigeneric

Describes a type of hybrid derived from a cross between two different genera. For instance *Heucherella* is a bigeneric hybrid between a heuchera and a tiarella.

Bipinnate

Term applied to leaves composed of several separate segments, which are themselves divided into separate segments.

Bloom

Usually taken to mean a flower, but can also mean a fine powdery coating.

Border

Generally taken to be a bed viewed from one side, like a traditional herbaceous border (as opposed to an island bed).

Although borders are often devoted to one type of plant—such as herbaceous plants or shrubs—there is no reason why you should not combine suitable shrubs and herbaceous plants, as well as bulbs and a seasonal display of annuals, in a mixed border.

Botrytis

Popularly known as grey mould, this fungus disease will not usually be a problem on established plants, but it may affect cuttings. The fungus usually starts on dead or dying tissue (dead or dying leaves or flowers for instance), but if left unchecked can spread to affect healthy tissue. The common name of this disease is apt—the affected part becomes enveloped in a grey mound of fluffy mould. If moved or disturbed, dust-like spores will probably fly up and drift to start new infections.

Many modern fungicides will control botrytis if you treat the plant early. It is worth dipping your cuttings in a solution of fungicide as a routine precaution.

Bract

A modified leaf at the base of a flower, sometimes highly coloured or the main feature of the plant (as in some euphorbias).

Bulb

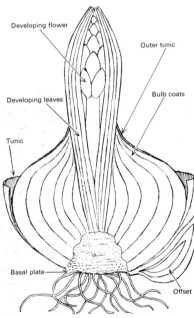

The term bulb has sometimes been used loosely in this book to indicate an underground storage organ that you might buy from a bulb merchant—such as a corm or tuber. A true bulb consists of fleshy or scale-like leaves arising from a base-plate. Usually, as with bulbous irises and daffodils, these scales are enclosed in an outer skin, but with lilies there is no outer cover.

Bulbils

Very small bulbs that form on some bulbous plants, which if detached will grow into full-sized bulbs. Bulbils form in the leaf axils of a few plants, such as the tiger lily.

Calcareous

Containing chalk or lime. A calcareous soil contains chalk or lime.

Calcifuge

A plant that dislikes lime or chalk, such as rhododendron.

Callus

Thickened tissue that forms over a wound.

Calyx (plural calyces)

The outer, protective part of a flower, consisting of a ring of usually green modified leaves (the sepals) that are fused together at the base to form a bowl, funnel, or tube.

Capsid bugs

There are several types of capsid bug, but they all cause similar damage. Both nymphs and adults feed on the sap of young shoots. The damaged leaves look puckered and distorted. Buds may be killed and flowers are often misshapen.

Capsid bugs are difficult to control as they tend to move about a lot, but frequent spraying with an insecticide may help.

Caterpillars

These common garden pests are the larval stage of various moths and butterflies. Although most feed on leaves, some live in the soil and feed on roots.

Caterpillars can do a lot of damage to vulnerable plants. They are best controlled by using a contact insecticide with some residual action. There is a lot you can do by hand picking if the infestation is small. Examining leaves, removing those with egg clusters if you notice them, might also be a way of eliminating most of the problem at source.

Cloche

A temporary structure (traditionally glass but increasingly plastic) for protecting plants in the open. Tent cloches are shaped like an inverted 'V', barn cloches have upright sides beneath the sloping roof panes. Tunnel cloches are usually made from a long continuous sheet of polythene (polyethylene) stretched over hoops and tensioned with wires.

Cloches can be used in the flower garden for hastening the germination of seeds sown directly in the open ground, and individual cloches are useful for protecting plants of borderline hardiness in winter.

Clone

A group of identical plants produced vegetatively from one original parent.

Chlorosis

The loss of, or lack of, chlorophyll (the green colouring matter in leaves) usually due to the lack of certain available elements in the compost necessary for its production. The leaves become bleached or yellow.

Plants that prefer an acid compost may become chlorotic if the compost contains too much lime, which prevents these plants from obtaining some essential nutrients.

Chlorotic plants can be treated with a chelated compound (Sequestrene).

Cold frame

Sometimes called garden frame. A structure offering weather protection—traditionally made from wood and glass but increasingly now from aluminium, metal, and plastics—though glass is still popular for the glazing.

Frames with glass sides have the advantage of better light penetration, but are colder in winter unless insulated then.

Frames are usually unheated, but soil or air warming cables can be used.

Unless otherwise stated, cuttings can be inserted directly into good soil or compost at the base. It is usual, however, to insert cuttings or sow seeds of border plants in containers such as pots or seed trays, rather than directly into the soil.

Frames have been suggested for propagating many plants in this book. But you will not succeed if the

Cold frame

plants are allowed to dry out—and those in containers are particularly vulnerable. *Plants in containers in a frame will need almost as much regular attention as those in a greenhouse.*

Compost
There are two meanings: *Garden compost* is the fibrous material, rich in humus, from the compost heap. *Potting compost* is a mixture of ingredients such as loam (soil), sand, peat, and fertilisers, for using in pots or seed trays. Earlier generations of gardeners used to have favourite recipes with ingredients that most of us would find practically impossible to obtain now: leaf mould, quality fibrous loam, crushed bones, decomposed sheep manure, mortar rubble and crushed brick, and even crushed potsherds, are examples. Fortunately, most plants will grow well in one of the modern standard potting composts, whether loam based or peat based. The important thing is to grow on young plants in a proper potting compost and not ordinary garden soil, if you want to give them a good start.

For cuttings, peat-sand composts, horticultural vermiculite and perlite are all effective, as nutrient content is less important than structure for rooting.

Container-grown plant
A plant grown and sold in a container, usually a rigid or flexible plastic pot. Container-grown plants suffer much less root disturbance than bare-root or balled plants (*See* Balled plant, Bare-root plant), transplant better, and enable you to plant at any time of the year if the ground is not frozen or water-logged.

Beware of *containerised* plants, plants potted up shortly before sale, as these will be little better than bare-root plants. Try lifting the plant by the stem—if established in the container it will not pull out of the compost. Containerised plants

are not likely to be a problem with herbaceous plants.

Most herbaceous plants are sold in pots.

Corolla

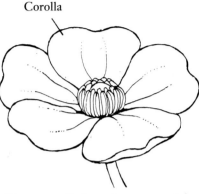

Corolla

Term applied to that part of the flower made up of true petals, usually forming a conspicuous inner whorl, backed or surrounded by the sepals, which comprise the calyx. While it is usual for the sepals to be green, occasionally they are brightly coloured and take over the role of the petals.

Cordate
Heart-shaped.

Corm
A corm is a flattened or thickened stem base, forming a storage organ. Unlike a bulb, a corm is solid and does not have distinct scales or layers if cut across, although there is an outer papery covering.

Corymb
Term used to describe a flat-topped cluster of flowers, composed of a series of flowers borne on individual stalks, each of different length. These form a rounded and more or less level head of flowers.

Crown
The basal part of a plant from which roots and shoots grow. Usually refers to the part of the plant just below or at soil level.

Cultivar
This is the botanist's name for a variety raised in cultivation. Some varieties of a plant occur naturally, and these are regarded as botanical varieties and should be printed in italics with a lower case initial letter, after the species name. Most varieties occur in cultivation (sometimes by chance, often the result of a breeding programme), and these are strictly 'cultivars', which should be printed in Roman type, in single quotation marks, and with a capital initial letter.

To the gardener it makes no difference whether the variety occurred in a garden or in the wild (it makes the plant no more or less desirable), and to most gardeners 'variety' is more familiar and less pedantic so we have used that term throughout this book.

Cutting
There are some special types of cuttings used for trees and shrubs, but herbaceous plants are propagated from simple soft cuttings, and most of these are prepared from basal shoots.

Root cuttings are also used for increasing some herbaceous plants and the preparation of these is described on the next page. *Soft stem cuttings* are prepared early in the season while the growth is still soft and succulent. Basal cuttings are taken from the new shoots arising from the crown of the plant in spring.

For most herbaceous plants the cuttings should be about 8 cm (3 in) long. Cut them off just below a node or joint, and trim off the lower leaves before inserting them in the compost.

A rooting hormone can be used, but just as important is humidity. Do not let the compost dry out, and mist the plants frequently if not in a propagator or enclosed environment (putting the pot in a polythene bag will help if you do not have a propagator, but do not let the bag come into contact with the cuttings).

Root cuttings are used for several herbaceous plants, such as anchusas, oriental poppies, gypsophilas, and verbascums.

Lift an established plant in late winter and select roots about the thickness of a pencil.

So that you know which end is which afterwards, make a straight cut across the top end and a sloping cut across the bottom.

Insert the cuttings so that the top is just below soil or compost level. When growth is evident, usually in spring, pot the plants up individually.

The roots of phlox are too thin to treat this way, so just lay them horizontally on the compost and cover lightly with compost or sandy soil.

Composts for cuttings. Use an all-purpose compost, or a seed compost (vermiculite and perlite are also very successful) for rooting cuttings; potting compost may contain a harmful level of nutrients.

A propagator will help to root the more difficult kinds, and mist propagation will help with the really tricky ones. You can, however, provide the necessary humidity with the aid of a polythene bag over the pot, but avoid direct sunlight, which could overheat or scorch the plants. *Shade* cuttings from direct sunlight until they have rooted and are growing strongly.

Dead-heading

The removal of faded flowers. This may be done to keep the plant tidy, to reduce the chance of self-sown seedlings becoming a nuisance, and sometimes to prolong the flowering period.

Deciduous

A plant that loses its leaves at the end of the growing season (in other words, it is not evergreen). It is normally used in the context of trees and shrubby plants, rather than herbaceous plants.

Deciduous trees and shrubs should be transplanted when dormant (unless container-grown), roughly from mid autumn to early spring.

Herbaceous plants are best planted or moved in spring or autumn.

Decumbent

Stems that lie on the ground for part of their length, then turn upwards.

Dentate

Coarsely-toothed, usually used to describe the edge of a leaf.

Digitate

Term applied to a leaf with finger-like, radiating leaflets.

Dioecious

Term used to indicate that all the flowers on any one plant are either entirely female or entirely male. Most skimmias are dioecious.

Disc (disk)

The central part of the flower, around which the petalled florets are arranged in the Compositae (daisy) family. The disc is composed of tiny tubular florets.

Dissected

Leaf or petal deeply cut into lobes or segments.

Division

Division is a self-explanatory term—it is simply a matter of dividing the old plant.

Not all plants are suitable. Only those with a fibrous root system and a crown that can be divided into sections containing shoots and roots can be propagated this way.

After lifting, many herbaceous perennials can be separated by hand, but those with fleshy crowns and roots may have to be cut with a knife.

Large clumps that cannot be handled easily can usually be forced apart by placing two garden forks back to back and putting pressure on the handles until the clump breaks up.

Dormant period

A temporary period when the plant ceases to grow, usually but not always coinciding with winter.

Drill

A straight, shallow furrow in which to sow seeds.

Earwigs

Earwigs damage the blooms of flowers such as chrysanthemums and dahlias, and may also make ragged holes in the petals and leaves of other plants.

Trapping (use an inverted flower pot filled with straw) will do much to control them if the traps are dealt with every day. Otherwise spray or dust around affected plants.

Entire

Term applied to a leaf which is undivided, with smooth margins.

Ericaceous compost

An acid compost containing very little or no lime. Ericaceous plants (those belonging to the Ericaceae or heather family) are generally lime-haters, but plants belonging to other families may also dislike lime in the compost.

You can buy both loam-based and peat-based ericaceous compost mixes, though they are less widely available than ordinary composts.

A lime-free compost should be used for raising seedlings or propagating cuttings of lime-hating plants.

Evergreen

A plant that does not shed its leaves all at the same time, like a deciduous tree or shrub. Although the leaves are shed eventually as new ones grow, this is done gradually so the plant always appears to be clothed.

Evergreen trees and shrubs are useful for their year-round interest, but too many can give the garden a 'heavy' appearance, and plenty of deciduous trees and shrubs and herbaceous plants are also needed to add variety and changing interest to the garden.

Some of the border plants in this book are evergreen—bergenias for instance. On this count some people would not regard them as true herbaceous plants, but in every other respect they belong in the herbaceous border. Although evergreen border plants can never make the herbaceous border very attractive in winter, they will help it to look less desolate.

Eye

There are two meanings. Applied to the flower markings it is a central area of a distinct colour.

It also means a dormant growth bud.

Fall

Pendulous outer petals of an iris.

Family

One of the groupings into which botanists place plants. All plants within a family will have some common characteristics that are not found in other families. The family name ends in 'ae': Liliaceae and Compositae for example.

Farinose

Covered with a white, waxy powder.

Feeding

Most of the plants in this book will benefit from a balanced general fertiliser raked in around them each spring.

It is a good idea to add bonemeal when planting.

Flore-pleno

A term that indicates doubleness, though the plant may be semi-double.

Floret

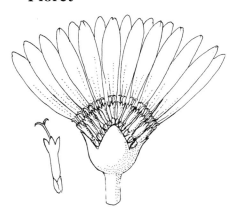

A single flower that forms part of a large head. In the daisy family (Compositae), each flower is really a head of tiny but closely-packed florets arranged around a central disc.

Foliar feeding

Some fertilisers are sold specifically as foliar feeds, but many ordinary liquid (and soluble powder) fertilisers can also be used as a foliar feed (the plants absorb the nutrients through the leaves). Generally it will do just as much good down at the roots, but sometimes an ailing plant or one that seems slow to grow away after planting may benefit from a foliar feed.

Frost protection

Most of the plants in this book are very hardy, but some of them are suitable for only very mild areas, while others will tolerate a moderately severe winter but not a very bad winter. You can help these plants by giving them a sheltered position, ideally near a wall that will protect them from the prevailing cold wind, and by making sure they have additional protection while young.

Some slightly vulnerable herbaceous plants may need no more protection than a covering of bracken, leaves, or peat for the winter. You may need to anchor this in place with a piece of netting.

A cloche may also be useful for individual plants.

Fungicide

A chemical that will kill or control fungus diseases (or at least some of them). Some have a systemic action (although it may not be as effective as spraying the leaves).

Gall

An abnormal outgrowth sometimes seen on trees and shrubs. A well-known one affects oaks, the so-called oak apples.

Galls are the result of irritation set up by insects or bacteria. Sometimes galls can be as big as a football, but they are usually more the size of a marble or smaller. They seem to have little or no effect on the plant.

Genus (plural genera)

A group of allied species. The genus is the first word of a plant name, and has a capital letter when written as part of a full name: in *Achillea filipendulina*, *Achillea* is the genus and *filipendulina* the species (*see* Species).

The genus is equivalent to a Surname, the species to the individual and unique member of the family. A genus may contain only one species, or it may contain more than a thousand. It depends on how closely botanists consider various plants are related.

Germination

The emergence of a new plant from a seed. *See* Seeds.

Glabrous

Smooth, though strictly it means hairless.

Glaucous

Bluish-grey; covered with a 'bloom'. Usually used to describe leaves or stems.

Grafting

Grafting is the joining of two plants so that they unite and grow as one plant. Grafting is sometimes used to control the plant's rate of growth (the rootstock often determining vigour), but grafting also provides a means of propagating those plants that are difficult to raise from cuttings or that will not come true to type from seed.

Grey mould

A disease characterised by a grey, fluffy growth of mould (*See* Botrytis).

Ground cover

An ornamental plant that itself requires little attention but covers the ground beneath it well enough to suppress weed growth.

Half-hardy

Likely to be damaged by frost. Usually applied to bedding plants that can spend the summer outdoors but not the winter. But some perennial border plants—such as *Lobelia cardinalis*—can be regarded as half-hardy in most districts.

Hardening off

The process of acclimatising a plant to cooler conditions. Plants frequently receive a severe check to growth if moved suddenly from a warm atmosphere into a cold, exposed one.

Cold frames are often used for hardening off greenhouse-raised plants, but plants raised in frames will themselves need hardening off. The frames are normally covered in cold weather, but as warmer conditions arrive the lights (tops) should be left off during the day and put back on at night. After a few weeks it should be possible to leave the lights off at night too.

Hardy

Generally taken to mean a plant that will not be killed by frost. But there are degrees of hardiness, and the more severe and prolonged the frost the greater the number of plants likely to suffer damage or be killed by it.

Heel

A small strip of bark and wood that remains attached to a shoot used for a cutting when it is pulled away from the main stem. Normally the heel is trimmed back to shorten it before the cutting is inserted in compost.

Cuttings that do not require a heel are normally cut off with a knife.

Heeled cuttings are not used for herbaceous plants.

Herbaceous border

In the strict sense, a border used only for herbaceous perennials. But many shrubs blend very well with normal herbaceous plants—*Perovskia atriplicifolia* and *Spiraea* x *bumalda* for instance.

Mixed borders, containing both shrubs and herbaceous plants, are becoming more popular.

Herbaceous plant

A plant that does not form permanent woody stems. That botanical definition would include annuals and biennials, but in common usage the term means perennials that die down in autumn and reappear in spring. But not all the plants in this book fit that description. Some, such as the bergenias, stachys and tiarellas retain their leaves through the winter.

Honeydew

A sticky secretion left on leaves by insects such as aphids and whitefly. It can be a particular problem because an unsightly black mould often grows on it, and this looks disfiguring.

The solution is simple—control the insects responsible for the honeydew secretion.

Hormone rooting preparations

Hormone rooting powders or liquids can be useful for rooting the more difficult plants, but many cuttings will root readily without assistance.

The two chemicals most widely used are naphthyl acetic acid (NAA) and indole butyric acid (IBA). There is some evidence that IBA is more effective on a wider range of plants, but both are useful.

Humus

The dark brown residue left when organic matter decays. The term is often used to describe the partly decayed brown, crumbly material such as well-made compost or thoroughly rotted leaves.

Hybrid

Usually a plant derived from crossing two distinct species or (much less commonly) genera.

Sometimes a cross *within* a species is also described as a hybrid. But unlike an ordinary cross resulting from normal cross-pollination, which might produce an ordinary variety (technically, cultivar), these hybrids are likely to be the result of crossing two pure 'lines' (in other words stable varieties that breed true). The resulting cross is called an F1 hybrid, and is generally better than the contributing parents. There are not many F1 hybrid border plants.

Inflorescence

The part of a plant bearing flowers. It is a term often used not so much in its botanical sense but to describe a flower head or spike that is unlike a typical flower (perhaps a spike where colourful bracts are the main feature, with the true flowers of secondary importance).

Indumentum

A dense covering of short hairs. The term is usually used to describe leaves or stems densely covered with short hairs.

Insecticide

A chemical for killing insects. There are many from which to choose, and most of them will be very effective against the insects that they are supposed to control. Sometimes, however, it is necessary to spray or dust more than once to achieve control. Always read and follow the manufacturer's instructions precisely.

Systemic insecticides are translocated within the plant and can be useful against insects that are difficult to reach or control with contact insecticides.

Internode

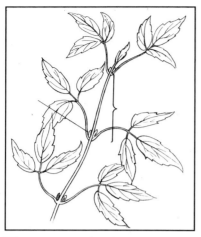

The portion of stem between two nodes (joints from which leaves arise).

Island bed

A bed intended to be viewed from all sides. Unlike a single-sided border with the tall plants at the back and dwarf ones at the front, an island bed has tall plants in the centre.

Juvenile foliage

Some trees and shrubs produce leaves of a different shape when they are young. Those of an ivy are less lobed when the plant has reached a mature, flowering stage. Leaves on a young eucalyptus are usually rounded and may clasp the stem, those on older trees are generally sickle-shaped and stalked.

Sometimes trees such as eucalyptus are coppiced or 'stooled' to stimulate a continuous supply of juvenile foliage.

Laciniate

Cut into narrow segments. Usually applied to leaves that are divided into fine segments.

Lanceolate

Term used to describe a lance-shaped leaf; one with a long, gradual taper.

Lateral shoot

Any shoot growing sideways from the main stem below the tip.

Layering

A method of propagation. Several types of ground and air layering are used for trees and shrubs, but only simple layering is described here. It is used mainly for propagating border carnations and pinks, which are layered in mid summer after they have flowered.

Cultivate the soil well where the plant is to be layered, incorporating peat and grit or coarse sand. Then trim the leaves and sideshoots off the chosen stem where it will be in contact with the ground. Use a sharp knife to slice horizontally into the stem, being careful not to cut through the stem.

Peg down the stem so that the slit area is buried. Use a piece of bent wire, or even an old-fashioned clothes peg, to hold the stem down. Cover the layered stem with fine soil, leaving the leafy tip exposed. Sprinkle sand around the layers. Water well and keep moist until well rooted.

When the plants have rooted, remove the pegs and sever the layer from the parent plant. Set out in a nursery bed to grow on.

Leader
The terminal shoot of a branch, which if left would continue to extend the line of growth.

A 'replacement leader' is a side growth that can take over the position of leading shoot once the leader has been cut out.

This term is used in connection with trees, not herbaceous plants.

Leaf bud cutting
Leaf bud cuttings are not much used by amateurs, but the technique can be useful for raising a lot of camellias for instance.

To make a leaf bud cutting on a suitable plant, cut just above and below the node, leaving a leaf attached. Pot up each leaf separately, just inserting the piece of stem into the compost.

Given warm, humid conditions the cuttings should root and begin to grow.

Leaf miners
The larvae of various moths and flies tunnel into the leaves of plants, sometimes leaving a winding white trail behind them, sometimes creating a white blotch.

Although disfiguring, it is usually possible to remove and burn affected leaves. It may be possible to kill the larvae in the tunnel by squeezing that part of the leaf between finger and thumb. If they become a real problem, spray the plants with a systemic insecticide.

Leaflet
A section of a compound leaf, usually resembling a leaf in itself.

Leaf mould
A confusing term. It has two meanings. It can be rotted leaves, an ingredient of some traditional potting compost mixtures; or a general term applied to various fungus diseases that rot leaves.

Most leaf-rotting diseases can be controlled to some extent by modern fungicides.

Leaf spot
The name given to several fungal and bacterial diseases that cause brown or discoloured patches on leaves.

Leggy
A term used to describe a plant that has become drawn and spindly, usually because of lack of light. The stem becomes elongated between the leaf joints, and the plant is generally weakened. It happens to plants in poor light, perhaps because they are shaded by overhanging trees or shrubs.

Linear
Long and narrow. Term used to describe leaves of this shape.

Loam
In many ways, the 'ideal' soil: neither wet and sticky nor dry and sandy. A good blend of clay, silt, sand, and humus. Good loam has a fibrous texture.

Loam is a basic ingredient of traditional potting composts, and is an ideal soil for the majority of trees and shrubs.

Loamless composts
Not all composts contain loam. Many are based on peat, but materials such as vermiculite may also be used. *See* Composts.

Lobe
Rounded segment that protrudes from the rest of the leaf or petal.

Mealy bug

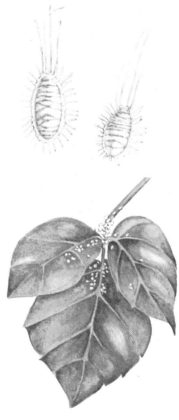

Easily identified pest. The young bugs are protected with what looks like a piece of cotton-wool. If only a few are present, pick them off. Otherwise use a systemic insecticide or spray with malathion.

Micropropagation
A scientific method of propagation, usually using very tiny portions of the growing tips (though there are several techniques) and growing these on using nutrient gels in

sterile conditions to provide a supply of plants. Sometimes micropropagation is used to produce a large number of plants commercially, sometimes as a method of providing healthy virus-free stock.

It is possible to buy the equipment and nutrients to do this at home, but unnecessary unless you want to try it for fun.

Midrib
The main rib that divides a leaf centrally along the length. It usually stands out more prominently on the back of the leaf than other veins and ribs.

Mildew

A white, powdery deposit on the leaves, sometimes spreading to the stem too. Some plants are more susceptible than others. Phlox and Michaelmas daisies are among the border plants that are notoriously vulnerable.

Pick off affected parts as soon as noticed. If the attack is severe it may respond to some of the modern fungicides, but the sooner an attack is treated the better the chances of success.

Mist
Misting a plant is a way of increasing humidity, which can be beneficial when rooting cuttings. It can be very successful *if done often enough*—even once a day may not be enough as the effect is fairly transient.

Use a fine mist, and do not spray plants with hairy leaves.

Mist propagation
Mist propagation is widely used commercially—it can drastically improve the 'take' of some of the more difficult plants being raised from cuttings. But if you need only a few plants it probably is not worth the expense of buying the equipment. And a high success rate is less important if only a few plants are needed.

A mist propagator prevents the cuttings wilting by keeping them covered with a fine film of water. It does this by intermittently spraying them with a fine mist. Remember that you need to install a water and power supply (the system is likely to have a soil-warming system too).

Most cuttings will root more quickly, with a higher success rate, but a mist propagator is most useful for the plants that are normally quite difficult.

Monocarpic
A plant that dies after flowering and seeding. Although annuals and biennials are monocarpic to this extent, the term is usually applied to perennials that grow for several years before flowering and dying.

Monoecious
A plant with separate male and female flowers *on the same plant* (but not in combined male/female flowers). The hazel is an example of a monoecious plant—the showy catkins are the males; the female catkins are small, red, and grow close to the stem.

Monotypic
A genus represented by only one species.

Mulch
A thick dressing of garden compost, peat, or similar material applied around the plants on the surface of the soil. Pulverised bark is another popular mulching material, but even black polythene or gravel can be used.

Mulching is particularly worthwhile during the first few years while herbaceous plants are becoming established, as it will conserve moisture and reduce weed competition.

Any organic material such as compost, peat, or pulverised bark will need to be at least 5 cm (2 in) thick to suppress weeds.

Naturalise
To grow in a simulated natural setting, such as in woodland or grass.

Neutral soil
A soil with a pH of about 6.5, neither very acid nor very alkaline. *See* pH.

Node
A point on the stem from which leaves arise. In the axils of the leaves are buds from which new shoots can develop.

The space between nodes (leaf joints) is known as the internode.

Nursery bed
An area of ground set aside in which to grow on young plants.

Oblong
Leaf with parallel sides, three times longer than wide.

Obovate

Egg-shaped in outline, broadest at the tip.

Offset

A young plant arising naturally on the parent and easily separated as an individual plant. Small bulbs and corms attached to the parent organ are also called offsets.

Opposite

Leaves that arise in pairs along the stem.

Orbicular

Leaves or petals that are disc-shaped, or almost so.

Ovate

Egg-shaped in outline, broadest at the base.

Palmate

A compound leaf shape, with lobes radiating like the fingers of a hand.

Peony wilt

Also known as peony blight, this disease can be a serious problem of tree and herbaceous peonies—especially in a wet season. The fungus attacks the stem bases of herbaceous peonies, showing itself as a grey mould, spreading to form brown blotches on the leaves.

Cut out and burn affected shoots, taking them back below soil level. Dust or spray the crowns with a fungicide, and do this the next year once the leaves appear, and fortnightly until the flower buds show.

Panicle

A branched flower cluster, each branch with numerous flowers on individual stalks, the youngest flowers at the top.

Pappus

Tuft of hair or bristles found in some flowers that later help the seeds to become airborne.

Peat compost

Peat-based composts have become very popular in recent years. They can be produced to a consistent standard, and are light and not unpleasant to handle.

Lime is added to bring the pH up to an acceptable level, and nutrients to support plant growth. Sand or other ingredients may be used to produce a suitable structure and texture.

Do not assume that a peat-based compost will be suitable for lime-hating plants (unless of course it is described as an ericaceous mix).

Peltate

A term used to describe a leaf which has the stalk attached to the centre of the underside (a nasturtium leaf for instance), and not to the edge of the leaf.

Perennial

A plant that lives for more than two years. Some plants that are perennial in the wild may be treated as annuals in cultivation.

Petiole

Leaf stalk.

pH

A scale by which the acidity or alkalinity of the soil or compost is measured. The scale runs from 0 to 14, though the extremes are never encountered in horticulture. Although 7 is technically neutral, 6.5 can be regarded as neutral horticulturally, as most plants will grow at this level. Most plants will tolerate a relatively wide pH range, but where particular trees or shrubs need an acid soil this has been indicated in the relevant entry in the first part of the book.

Phototropism

There is a propensity for plants to grow towards light, which can be a particular problem in a very shady area, where light is usually from one direction.

Plants show varying degrees of phototropism.

Pinching out

This is a form of pruning to encourage the plants to branch out. By removing the growing tip, shoots below are encouraged to branch out to take over. It is a technique usually used for plants that would otherwise tend to become leggy, and always when the plant is relatively young.

Pinna (plural pinnae)

The individual leaflet of a deeply divided leaf, such as most fern fronds.

Piping

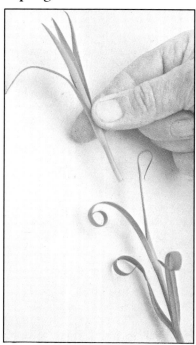

A cutting obtained by pulling off the tip of a non-flowering shoot of a carnation or pink.

Planting

It is worth taking time and care over planting as you only have one chance to get things right.

Dig a hole large enough to spread the root system fully, and mix a bucketful of peat and a handful of bonemeal into the excavated soil (which will be returned later).

Loosen any compacted ground at the bottom of the hole, and if the soil is poor or impoverished, lightly fork in some peat and bonemeal.

Set the plant in the hole to the same depth as it was before.

If planting a pot-grown herbaceous plant, tease some of the roots out if they have started to run around the bottom or edge of the container.

If bare-rooted, trickle some soil among the roots first. Then add more soil, firm well, and finish off by levelling and then loosening the surface to avoid compaction.

Plunge

Act of burying a pot up to its rim in the ground. This makes it less liable to dry out and makes watering less critical.

Instead of plunging the pots into the ground, an area can be boarded off and filled with peat or grit for the same purpose.

Pot-bound

A term used to indicate the stage at which the pot has become so full of roots that growth of the plant is suffering. Most plants need re-potting once the compost appears full of roots.

Pot-grown plant

A plant bought in a pot. For larger plants such as shrubs the term 'container-grown' is more usual, but young herbaceous plants in small pots may be referred to as 'pot-grown'.

Pots

Pots now come in many shapes and sizes—including square—not to mention colours.

Most pots nowadays are plastic, and for many plants these are perfectly suitable. The compost is less likely to dry out so quickly, and they wipe clean easily so unsightly deposits are unusual.

Be cautious of colours—some will be affected by strong sunlight.

Do not dismiss clay pots. They will make overwatering less likely, and for plants with large or heavy top growth the extra weight might be a useful counterbalance in windy weather.

Potting on

Moving a plant on into a pot of a larger size, usually one or two sizes up. Never pot on into a pot much larger than the present size, otherwise the compost may become either exhausted or 'sour' before the plant has chance to use it.

Pot up

To move a seedling or rooted cutting into a pot for the first time.

Prechill

Seeds of many perennials are much slower and more difficult to germinate than, say, annuals or most bedding plants. It is often necessary to break their dormancy or to remove germination inhibitors.

Traditionally this was done by sowing them in autumn or early winter and subjecting them to a period of cold outdoors before trying to germinate them, and this has been recommended for many of the plants described in this book.

Nature can be speeded up another way with the aid of a refrigerator.

Place the seeds between *moist* blotting paper, thickly folded kitchen roll, or a piece of flannel, in a container with a lid. Leave this in the refrigerator for two or three weeks, making sure the seeds are kept moist but not waterlogged, then sow in warmth.

Prechilling can be tried for all those plants for which a cold period has been suggested.

Presoaking

Some seeds will germinate more readily if soaked for 12 hours before sowing. Soaking in tepid water has been suggested for a few plants in this book. Obviously without special facilities it is difficult to keep the water tepid for that length of time, but no harm will come if you simply change the water as often as possible during that time.

Pricking out

A term used to describe the job of lifting individual seedlings from the box or pot in which they were sown and planting them into individual pots or spacing them out in a box.

Propagation

Act of raising new plants by seed or vegetative means (cuttings, layering, for instance).

Propagator

Cabinet for germinating seeds, raising seedlings, or rooting cuttings. Most are heated to provide the necessary warmth, and enclosed to maintain a very humid atmosphere.

Remember that a propagator must be placed in good light (though certain seeds may germinate better in darkness, the seedlings will need a light position). Avoid direct sunlight, however, as this may increase the temperature too much and cause scorching.

Raceme

An elongated unbranched flower cluster, the individual flowers being stalked.

Repotting

The term repotting is loosely used to mean putting the plant into a new pot, whether the same size or larger. More specifically it means replanting in the same-sized pot after removing some of the old compost and replacing with new. Moving the plant on to a larger pot is more correctly 'potting on'.

Rest/resting period

Many plants have a natural resting period, when they are either dormant or making little or no new growth. This is not necessarily accompanied by a loss of foliage.

In temperate climates many plants rest in the winter, although a number of bulbs are exceptions to this rule. Crocuses, tulips and daffodils, for example, rest during the summer after they have flowered.

Rhizome

A horizontally creeping underground stem that acts as a storage organ.

Rootball

The mass of roots and compost when a plant is removed from its pot.

Rosette

An arrangement of clustered leaves radiating in a whorl from a central point.

Runner

A long shoot sent out by some plants, that will root and form new plants where it comes into contact with the soil.

Scale

Scale insects look like small yellowish-brown insects that resemble small scales or shells. They are immobile. You can wipe them off with a sponge soaked with an insecticide (but use waterproof gloves). You can also try a systemic insecticide.

Seeds, plants from

Most herbaceous plants can be raised from seed, though it may be a fairly slow process. It does, however, solve the problem if you do not already have a plant that you can divide or from which you can take cuttings. Most general seedsmen offer a reasonable range of herbaceous perennials and bulbs, but you may have to go to specialist seedsmen for most of them.

Hybrids and varieties of most perennials are unlikely to come true from seed, and the method is most suitable for the species, or where you do not mind if the plants are variable.

Be prepared for disappointments if you try seed of highly bred and selected plants such as border phlox and Michaelmas daisies. These are most unlikely to be comparable with the named varieties that are propagated vegetatively.

On the other hand a packet of Russell lupin seeds is likely to produce results as good as any that you buy as plants.

If only a few plants are needed, sow the seeds thinly in a pot of seed compost. If a number are required, then prepare a seed tray as shown above.

Level the surface of the compost, firming it gently, and space the seeds well. Cover with a sprinkling of compost.

Water the seed tray, or place the pot in a bowl of water to let moisture soak up from beneath.

Then cover with a piece of glass, if you do not have a propagator, and a sheet of paper to exclude light.

Once the seedlings have germinated, prick them out into another seed tray or individual pots.

Semi-ripe
Term used to describe a type of shoot used for tree and shrub cuttings. These are the current season's shoots in late summer, when growth has slowed down and the wood started to mature.

Serrated
Toothed, like a saw. A term usually applied to the margin of a leaf.

Shrub
A woody plant without a single tree-like trunk. Some shrubs can be grown as trees, depending on how they are trained.

Slugs and snails
Slugs and snails can be a menace as the succulent young shoots of herbaceous plants are emerging in spring. Slug baits are very effective—if you are worried about pets or wildlife eating the pellets, simply place them beneath stones or pieces of slate. Generally, however, mini-pellets of bait coloured blue are not likely to be picked up by animals or birds.

Species
An individual member of a genus. *See* Genus.

Specimen plant
A tree or shrub (usually) planted to be viewed from all angles, not as part of a large group.

Spike
Strictly an unbranched flower stem with stalkless flowers arranged around it. Sometimes loosely used to indicate any flower head with a spiky or poker-like appearance.

Staking
It is a good idea to keep the number of plants that need staking down to a minimum, but there will be good plants that you will want to grow that simply have to be staked if they are not to sprawl or collapse in high winds.

Start staking *early*—it will be far more effective, and because the plants can grow through the support it will be far less conspicuous.

There are several proprietary plant supports that you will see at garden centres or advertised in gardening magazines. These can be useful if positioned early enough, but they are unlikely to be as effective as bushy twigs—what gardeners traditionally call 'pea sticks'. These give excellent support and are unobtrusive once the plants have grown.

Stamen
Male reproductive organ, which forms and carries the pollen.

Stool
There are two uses of the word. A tree or shrub when grown as a cluster of young stems arising from ground level, achieved by cutting the shoots back close to the ground annually.

The term is also used to describe the crown of an herbaceous plant that is lifted annually for propagation.

Stop/stopping
To 'stop' a plant is to pinch out the growing tip to encourage lateral growths to develop. *See* Pinching out.

Stratify

To break dormancy of seeds by exposing them to low temperatures before sowing. Fleshy fruits are crushed first, then placed in layers of sand or peat in pots or boxes and stood outdoors (protected from vermin) during the winter. The seeds are cleaned and then sown in warmth. *See also* Prechilling.

Strike

To root a cutting.

Systemic

A term used to describe insecticides and fungicides that can be taken up by the plant and translocated (moved about) within the plant. The chemical can be taken up by the roots and will find its way to the leaves.

Tender

A term indicating a plant likely to be damaged by frost.

Tendril

A thread-like modified leaf or stem that can twine round a support to enable the plant to cling and climb.

Tomentose

Densely covered with fine hairs.

Tree

A woody plant with a distinct trunk or main stem, though some trees have several trunks. Some trees can be grown as shrubs, depending on the initial training.

Trifoliate

A leaf divided into three leaflets.

Tuber

A thickened fleshy root (as in a dahlia) or a swollen underground stem (such as a potato). The size and shape of tubers vary considerably, but they are all a means of survival in periods of drought or cold.

Tunic

A dry, often papery, skin covering corms and some bulbs.

Umbel

A cluster of flowers in which all the individual stalks arise from a common point at the top of the main flower stem.

A compound umbel is an umbel of umbels.

Variegated

Leaves that are spotted, blotched, edged, or otherwise marked with another colour.

Vegetative propagation

Any method of increasing a plant other than by seed.

Virus diseases

These are impossible to control in the garden in any practical way, and affected plants are best destroyed (except of course those few plants grown for their decorative variegation caused by viruses).

Symptoms are generally mottled, yellowing or distorted leaves, but first make sure that the problem is not caused by insects such as greenfly or by a nutritional deficiency.

If the plant is also stunted, it is best to destroy it without delay— even if it is not due to a virus it could be a problem such as eelworm, which is also difficult to control.

Virus diseases are usually transmitted by aphids and other sucking insects, so controlling these will reduce the chances of infection.

Whorl

An arrangement of leaves or flowers arising from one point, arranged rather like the spokes of a wheel.

Common Name Index

The most important Latin synonyms are also included

Picture credits
Bernard Alfieri: 78(t)
H. Allen: 113, 118(br)
Amateur Gardening: 57, 61, 107(br)
D. Arminson: 86, 95(b), 103(t)
P. Ayers: 118(bl)
R.J. Corbin: 56(b), 67, 78(b), 79(t), 81, 83,
  106(tr), 110(tr, br), 111, 114(tc, tr, br)
J.E. Downward: 36
Alan Duns: 14(t), 17(b), 21, 62(t), 94(t)
Valerie Finnis: 13, 17(t), 20(b), 27, 42(t), 44,
  46(t), 59, 69(t), 73, 74(t), 88(b)
P. Genereux: 28
P. Hunt: 9, 20(t), 50(t), 58, 65(tr), 71
A.J. Huxley: 22(b), 56(t)
George Hyde: 24(t), 107(tr), 114(tl)

Leslie Johns: 30, 72(t)
R. Kaye: 82(b)
K. Knowles: 74(b)
E. Megson: 79(b), 90(t), 96(t)
Peter McHoy: 6, 10(b), 16, 18(t, b), 26(t),
  34(t), 38(b), 66(b), 90(b), 97,
  106(tl, bl, br), 107(bl), 109, 115(tr),
  117(tr)
Natural History Photographic Agency: 104
M. Newton: 7(t), 80(t)
Maurice Nimmo: 41
Opera Mundi: 40(t)
S.J. Orme: 110(bl)
R. Parrett: 68(t, b)
N. Procter: 107(tl)
Miki Slingsby: 31, 37(bl, br)

Harry Smith Horticultural Photographic
  Collection: 1, 2/3, 4, 5, 7(b), 8(t, b), 10(t),
  11, 12(t, b), 14(b), 15, 19, 22(t), 23, 24(b),
  25, 26(b), 29, 32(t, b), 33, 34(b), 38(t), 39,
  40(b), 42(b), 43(t, b), 45(t, b), 46(b), 47,
  48(t, b), 49, 50(b), 51, 52(t, b), 54(t, b),
  60(t, b), 62(b), 63, 64, 65(tl), 66(t), 69(b),
  70(t, b), 72(b), 75, 76, 77(t, b), 82(t),
  87(t, b), 89, 91, 92(t, b), 93, 94(b), 95(t),
  96(b), 98(t, b), 99, 100(t, b), 101, 103(b),
  118(bc)
G.S. Thomas: 35
Tourist Photo Library: 53, 80(b), 84(t)
Colin Watmough: 55
D. Woodland: 88(t)